Abel Joel Grout

The Bryologist

Abel Joel Grout

The Bryologist

ISBN/EAN: 9783741107719

Manufactured in Europe, USA, Canada, Australia, Japa

Cover: Foto ©berggeist007 / pixelio.de

Manufactured and distributed by brebook publishing software
(www.brebook.com)

Abel Joel Grout

The Bryologist

THE BRYOLOGIST

AN ILLUSTRATED BIMONTHLY

DEVOTED TO

NORTH AMERICAN MOSSES

HEPATICS AND LICHENS

VOLUME VII. 1904

EDITORS

ABEL JOEL GROUT AND ANNIE MORRILL SMITH

PUBLISHED BY THE EDITORS

78 ORANGE STREET, BROOKLYN, N. Y.

INDEX 1904

* Star indicates illustrations.

AUTHOR'S INDEX.

VOLUME VII.   NUMBER I

✸ JANUARY, 1904 ✸

THE BRYOLOGIST

AN ILLUSTRATED BIMONTHLY DEVOTED TO

NORTH AMERICAN MOSSES

HEPATICS AND LICHENS

EDITORS:

ABEL JOEL GROUT and ANNIE MORRILL SMITH

CONTENTS

Entered at the Post Office at Brooklyn, N. Y., April 2, 1900, as second class of mail
matter, under Act of March 3, 1879.

Published by the Editors, 78 Orange St., Brooklyn, N. Y., U. S. A.

PRESS OF MC BRIDE & STERN, 97-99 CLIFF STREET. NEW YORK

THE BRYOLOGIST

BIMONTHLY JOURNAL DEVOTED TO THE STUDY OF NORTH AMERICAN MOSSES
HEPATICS AND LICHENS.

ALSO OFFICIAL ORGAN
OF THE SULLIVANT MOSS CHAPTER OF THE AGASSIZ ASSOCIATION.

Subscription Price, $1.00 a year. 20c. a copy. Four issues 1898, 35c. Four issues 1899, 35c. Together, eight issues, 50c. Four issues 1900, 50c. Four issues 1901, 50c. Four Vols. $1.50 Six issues 1902, $1.00. Six issues 1903, $1.00.

Short articles and notes on mosses solicited from all students of the mosses. Address manuscript to A. J. Grout, Boys' High School, Brooklyn, N. Y. Address all inquiries and subscriptions to Mrs. Annie Morrill Smith, 78 Orange Street, Brooklyn, N. Y. For advertising space address Mrs. Smith. Check, except N. Y. City, MUST contain 10 cents extra for Clearing House charges.

Copyrighted 1904, by Annie Morrill Smith.

THE SULLIVANT MOSS CHAPTER.

President, Prof. J. M. Holzinger, Winona, Minn. Vice-President, Mrs. C. W. Harris, 125 St. Marks Avenue, Brooklyn, N. Y. Secretary, Miss Mary F. Miller. 1109 M Street, Washington, D. C. Treasurer, Mrs. Smith, 78 Orange Street, Brooklyn, N. Y.
Dues $1.10 a year, this includes a subscription to THE BRYOLOGIST.
All interested in the study of Mosses, Hepatics, and Lichens by correspondence are invited to join. Send dues direct to the Treasurer. For further information address the Secretary.

--------------- *The* ---------------

Journal of the Maine Ornithological Society

With the January, 1904, Number
Begins its Seventh Volume

A Quarterly Journal all about Maine Birds

You should have Mr. Arthur H. Norton's papers on "The Finches Found in Maine" and the series of papers on "The Warblers Found in Maine," written up by four of the members of the Society. These alone well worth the price of a year's subscription.

Subscription, Fifty Cents per Annum ❧ ❧ **Fifteen Cents per Copy**

Send stamps for Sample Copy to

J. MERTON SWAIN, Business Manager ❧ ❧ **FAIRFIELD, MAINE**

ALL ABOUT THE WILD-FLOWERS

THERE is only one publication *about* the wild-flowers and that is *The American Botanist*. It does not publish technical articles, and uses the common names of plants whenever possible. Full of notes on the haunts, habits, fragrance, uses, and products of plants. Every plant-lover should have it. Monthly, $1.00 a year. Sample for a 2-cent stamp.

ADDRESS, **THE AMERICAN BOTANIST** ❧ **Binghamton, N. Y.**

PLATE I. *Peltigera canina.*

THE BRYOLOGIST.

Vol. VII.   January, 1904.   No. 1.

LICHENS—PELTIGERA.

Carolyn W. Harris.

The genus Peltigera is one of the most common among the lichens.
While its range is generally northern, some species are found in all parts of
our globe where conditions are favorable to their growth. In Europe and
North America there are several species which are very abundant. Some
of the Peltigeras resemble closely some species of Nephroma, but the fruit
is borne on the upper side of the extended lobes of the thallus instead of on
the under side as in Nephroma.

It is very difficult to determine the different species when sterile because
of the great similarity of the thallus.

A well known lichenist says "it is impossible to do so with certainty."

The thallus is large, with rounded lobes, quite thick and soft when fresh,
becoming brittle when dried. The color is bluish or brownish green, which
changes to brown or brownish gray when dried. The lower surface is light
brown, with conspicuous veining, and has usually long, hair-like rhizoids
which cling so closely to the substratum, especially when growing on moss,
that it is difficult to separate them.

The apothecia are usually large and are terminal upon the extended
lobes of the thallus. When young they are peltate, but in many species are
revolute when fully developed.

The different species of Peltigera are found on damp, mossy rocks and
on the earth, often growing with mosses. They require a great deal of
moisture; when dry they shrivel and become very brittle. After a rain
when the cells are filled with water they unfold like a flower and are more
than doubled in size.

Peltigeras are supposed to be of rather slow growth and probably they
are until they are well started, after that they increase rapidly. One I
transplanted a year ago is now two thirds larger than it was at that time and
is fruiting abundantly.

No doubt many of the lichens which have been generally considered
of slow growth have not been carefully observed for any length of time.
Any one interested in the lichens having the opportunity of studying one
particular specimen throughout an entire year, or even longer, and carefully
keeping an account of its growth, fruiting, any change in form or color, would
be doing an educational work.

Peltigera venosa (L.) Hoffm. Thallus small, erect, ovate to fan

The November Bryologist was issued November 2d, 1903.

shaped, with broad, flat lobes on whose edges are borne the apothecia. The thallus is covered with a powdery bloom, in color it is a gray green, frequently turning a reddish brown when pressed. The under surface is nearly white, with dark brown felt-like branching veins traversing the surface of the thallus and growing into a thick mat at the center.

The apothecia are large, round and flat, and are found on the slightly elongated lobes of the upper side of the thallus. The disk is dark brown or black, orbicular, the margin thick with finely serrate edges.

P. VENOSA is a small, but handsome species; it is rare in the Eastern States, but quite common in the far west, grows on wet, clayey, or argillaceous soil, often in the crevices of rocks near waterfalls.

PELTIGERA APHTHOSA (L.)Hoffm. Fig. 1. The thallus of this species is large, smooth, rather soft, and covered with scattered appressed dark, irregularly shaped warts. The lobes are broad and round. The color an apple green turning to a dull green; when dried is usually a light yellow brown. The under surface is cream color, with reticulated blackish veins, which at the center form a close black nap.

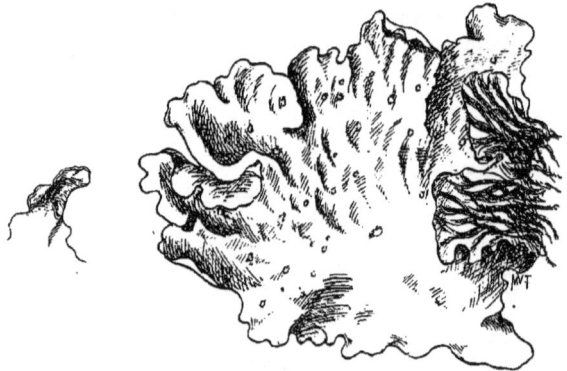

Fig. 1. *Peltigera aphthosa* × o.

The apothecia are large and round, borne on the ascending lobes of the thallus; when developed they become revolute and curl around these extended lobes.

P. aphthosa is a large and common species, is found on rocks among mosses and on the earth. It is easily distinguished from the other Peltigeras by the wart-like depressions on the upper side; in some specimens these are quite large and crowded, in others they are small and scattered.

PELTIGERA HORIZONTALIS (L.) Hoffm. This species resembles both *P. canina* and *P. polydactyla*, but the thallus is smooth, not downy as in *P. canina*, nor polished as in *P. polydactyla*. The thallus is large, in color is greenish to reddish brown. The under side is about the same color, is covered with dark veins which soon coalesce and form a dark brown felt-like nap. The fibrils are few and short, and almost black.

The apothecia are on the scarcely elongated lobes; they are oblong, horizontal, rather large; the disk is reddish brown, with a finely serrated margin.

P. horizontalis is not so common a species as *P. canina* and is found in the same localities; it is usually fertile.

PELTIGERA POLYDAC-
TYLA (Neck) Hoffm.
Fig. 2. Thallus large,
very smooth and thin,
with a somewhat pol-
ished surface; the lobes
are broad and round
with slightly crisped
margins. The color is a
bright grass green,
becoming lead color or
brown when pressed.
The under side is light
brown, with tufts of
dark brown fibrils, the
reticulated veins are a
light brown, turning
much darker toward the
center.
   The apothecia are
usually large and revo-
lute, digitately clustered

Fig. 2.   *Peltigera polydactyla* ✕ o.

and are borne on the extended lobes of the thallus. The disk is reddish
brown, with a crenate margin of lighter color. The habitat of this species
is rocks and decaying wood; it is a common one throughout the United
States.
   PELTIGERA SCUTATA (Dicks.) Leight. The thallus of *P. scutata* is small,
quite thin and somewhat rough; the color is a dull green, becoming a light
reddish brown. The lobes are narrow and crisped, with frequently gray
soredia on the margins. The under side is almost white with light brown
veins which forms a spongy nap.
   The apothecia are small and are found on the short and somewhat scattered lobes of the thallus; the disk is a dark, dull brown.
   P. scutata is found on rocks and earth with mosses. It is not a rare
species, but not so common as many others of the Peltigeras.
   PELTIGERA PULVERULENTA (Tayl.) Nyl. Thallus not as large as in *P.
polydactyla*, it is furrowed and pitted, thin and somewhat granulated, has
short, crisped lobes. In color is a grayish green, soon turning to a greenish
brown. The under side is light brown, with veins of the same color which
toward the center coalesce, forming a close nap.
   The fertile lobes are narrow and compressed, on these are borne the
medium sized apothecia, which are round, with dark brown disks: fruiting
specimens are rare.
   P. pulverulenta is found in both the Eastern and Western States, but
is not a common species; it grows on rocks and earth in very moist localties.
   PELTIGERA RUFESCENS (Neck.) Hoffm. This species so closely resembles

P. canina that they are often confused. The thallus of *P. rufescens* is not so downy as *P. canina*, it is more rigid and the lobes are narrower and more crowded, they are elevated and crisped. In color it is a greenish gray, becoming a dark reddish brown. The under surface has brown reticulated veins and dark brown fibrils.

The apothecia are usually large, oblong and finally revolute; the disk is a dark chestnut. This species, like most of the Peltigeras, is found on rocks and earth with mosses.

PELTIGERA CANINA (L,) Hoffm, Plate I. Thallus rather large, sometimes very large, is thin and tough, but soft and limp, is downy and often furrowed. In some specimens the thallus is smooth, except at the margins of the flat, rounded lobes, where it is wavy. In color it is grayish or brownish green. The under surface is a very light brown, almost white, with vertical veins and fibrils of the same color: these are short and thick giving a sponge like appearance.

The apothecia are quite large, at first are round and flat, but soon are somewhat revolute and elongated.

P. canina is a common species, is found on moist earth and on rocks. When pressed its color changes more than that of any other species of Peltigera; it loses all traces of green and becomes a reddish or russet brown. It is a widely distributed species, growing in Europe from Lapland to Switzerland, and throughout North and South America.

PELTIGERA CANINA (L.) Hoffm. var. SPONGIOSA Tuck. This variety, which is subalpine, corresponds to *P. canina* in general characteristics, the thallus is thinner, more brittle, and the under surface is covered with a close nap of white fibrils, which changes toward the center to darker reticulated veining. In some specimens these nap-like fibrils are quite long. A large mat of *P. spongiosa* was collected by Dr. and Mrs. N. L. Britton at Rifle Notch, Essex Co., New York, which when fresh had fibrils a quarter of an inch long. The upper side of the thallus was a delicate gray green, the lobes were long and broad.

POGONATUM URNIGERUM (L.) BEAUV.

MARY F. MILLER.

There is a railroad embankment near Shandaken, N. Y., where this moss is very abundant. This summer, on the 14th of July, I found thousands of the young sporophytes just making their appearance (Fig. 1). The blackened setae and capsules of the previous year were still attached to many of the plants. It was nearly two months, Sept. 9th, before the capsules were matured (Fig. 2). While collecting some of these fertile plants I noticed that a number of calyptras seemed to be turned inside out, and to stand straight up from the tips of the opercula. Examining the plants, I found calyptras in all stages of departure. It seems as though the calyptra makes ready to depart by an upward movement of all except its top (Fig. 3), that seems to sink in, or rather, the upper portion of the calyptra is so bunched

PLATE II. *Pogonatum urnigerum.*

up around it, that it gives it that appearance (Fig. 4.); and so the process goes on, until finally the calyptra is turned completely inside out (Fig. 5.). It is remarkable how long these frail things remain in this last position. I kept some of the plants moist four or five days, and even handled them somewhat roughly, before the calyptras fell off. When dried, they fall immediately.   Washington, D. C.

NOTES ON VERMONT MOSSES.

A. J. GROUT.

I. Recent Additions to the List of Vermont Mosses.

The following list of species new to Vermont is based mostly on my collections made during the summers of 1902 and 1903. A few of the additions are specimens that have been some time in my herbarium but which have only recently been determined. A critical study of a portion of Frost's collection, while preparing Part II of my "Mosses with Hand-Lens and Microscope," resulted in two or three finds, and doubtless there are more to be made when the whole collection shall be carefully studied.

This is the third additional list since the original list was published in 1898. The second addition was published in Rhodora for Sept. 1902. I am greatly indebted to Dr. G. N. Best for assistance in determining some of the most difficult specimens. In fact Dr. Best facetiously told me at one time that if there were more specimens in Vermont like those I had been sending him, I would better move to some other state.

BUXBAUMIA INDUSIATA Brid. Willoughby. Miss Annie Lorenz. Det. Mrs. Britton.

POLYTRICHUM SMITHIAE Grout, BRYOLOGIST, May, 1903. Mt. Mansfield.

DICRANODONTIUM LONGIROSTRE (W. &. M.) B. & S. Willoughby. Two stations. Miss Annie Lorenz.

DICRANUM BONJEANI DeNot. Old roofs, Newfane; swamp, Charlotte, Pringle.

DITRICHUM HOMOMALLUM (Hedw.) Hampe var. Det. Dr. Best. Clefts in rocks in cool ravine, Newfane.

BARBULA GRACILIS (Schleich.) Schwaegr. Crevices of limestone rocks, Brattleboro. Frost. Fide L. & J. Manual, p. 127.

TORTULA RURALIS (L.) Ehrh. Dry limestone rocks at base of cliffs, No. Pownal, Vt. A. LeRoy Andrews. Determined as *Tortula montana* by Dr. Best, but the leaf cells are fully 0.013 mm. in the upper portion so that it seems almost certain that the plants are depauperate specimens of the polymorphous *T. ruralis*.

GRIMMIA AMBIGUA Sulliv. Frost's *G. Donii* var. *sudetica* proves to be almost typical *G. ambigua*.

G. APOCARPA GRACILIS (Schleich.) Web. & Mohr. Cliffs, Glebe Mt., Newfane.

G. APOCARPA RIVULARIS (Brid.) Web. & Mohr. Common.

G. CONFERTA Funck. var. Limestone ledges, Stratton.

G. CONFERTA OBTUSIFOLIA Schimp. Dry limestone cliffs, Snake Mt.

G. PENNSYLVANICA Schwaegr. Dry ledges, Jamaica.

G. Pennsylvanica Bestii n. var. Growing in short, wide, loose tufts on stone in walls and on ledges in open fields. The stems are less than 5 mm. high; the leaves are shorter with shorter hair points and frequently with lamina of a single layer throughout. The lower leaf cells are characteristically sinuose as in the species. The leaves bear on the back numerous bodies which look like propagula, but which Mrs. Britton and Dr. Howe believe to be a species of Alga. This form is very common in Newfane and the surrounding towns, and for several years I have been attempting to locate it. Very likely it will prove to be a good species and if so it should be called *G. Bestii*.

ORTHOTRICHUM OBTUSIFOLIUM Schrad. On the bark of maple and pear trees, Newfane.

O. SCHIMPERI Hamm. On maple trees, Brandon. Miss Harriet Wheeler.

O. SPECIOSUM Nees. Frequent on fruit trees, Newfane.

BRYUM CAPILLARE L. On rocks, Brattleboro, Frost; Guilford, Mrs. J. B. Clapp.

AMBLYSTEGIUM JURATZKANUM Schimp. In an old cellar hole. Alt. 1,600 ft. Newfane. Det. L. S. Cheney.

FABROLESKEA AUSTINII (Sulliv.) Best. On pear tree with various species of *Orthrotrichum, Hypnum reptile* and *Pylaisia Schimperi.* The species covered about a square inch, and would have been overlooked with ordinary scrutiny but the tree on which it grew was carefully studied from my notes on moss habitats. It was distinguished from *Pylaisia* because the leaves were not homomallous, and from *Hypnum reptile* by the straight capsules. Det. Dr. G. N. Best.

HETEROCLADIUM SQUARROSULUM (Voit.) Lindb. Det. Dr. Best. Abundant and fruiting on humus over moist stones, Ball Mt., Townshend. Not before reported from eastern United States except Mt. Washington, and has been collected in fruit but twice before in North America. Issued as No. 169 of my N. A. Musci Pleurocarpi but wrongly identified as *Thuidium microphyllum.*

LESKEA NERVOSA NIGRESCENS (Kindb.) Best. Abundant on bases of sugar maple and other rough barked hardwoods in Newfane. No. 175, N. A. Musci Pleurocarpi.

NECKERA COMPLANATA (L.) Hueben. Willoughby. Miss Annie Lorenz. Det. Mrs. Britton.

II. Notes on Species Previously Listed.

ATRICHUM CRISPUM James. I feel sure that this is a mistake in Frost's list for there is a specimen of this species in his collection, but it is from New Jersey. I find no specimens of this species so far inland.

GRIMMIA UNICOLOR. In fruit. Brattleboro. Frost.

BARBULA CONVOLUTA. Newport. Faxon.

TRICHOSTOMUM CYLINDRICUM (Bruch.) C. M. (*Didymodon cylindricus* B. & S.). No Vt. specimens can be found. The specimen listed from Rock Point proves to be *Tortella tortuosa* (L.) Limpr. (*Barbula tortuosa* Web. & Mohr.).

BRYUM DUVALLII. Swampy soil. Alt. 1,600 feet, Newfane.

MNIUM HORNUM. Abundant along Clear Brook, Dover.

LESKEA NERVOSA. A depauperate form with numerous flagella at the ends of some of the branches, is very common about the bases of trees in Newfane.

GYMNOSTOMUM RUPESTRE Schleich. The only Vermont specimen correctly referred to this species is the one collected by Faxon at the foot of Mt. Hore cliffs, June 23, 1884. Most of the other specimens referred to this species are *G. curvirostrum* (Ehrh.) Hedw. var. *scabrum* Lindb. which has been collected at Rock Point, Burlington, by myself, and much earlier by Mr. Pringle on the "Cliffs of Lake Champlain." This var. is readily recognized by its papillose stems. *G. curvirostrum* has dark red-brown capsules, *G. rupestre* pale yellowish brown capsules.

THE GENUS HYMENOSTOMUM IN NORTH AMERICA.

JOHN M. HOLZINGER.

Have we this genus in North America? This depends somewhat on the point of view. I think we have here a moss which in Europe passes for a Hymenostomum.

HYMENOSTOMUM R. Br. in Trans. Linn. Soc., 1819. The following is the historical view of this genus drawn up by Prof. Limpricht in Laubmoose, I, p. 223, 224: " *Hymenostomum* (membrane-mouth or-orifice), founded on *Gymnostosmum microstomum* Hedw., is characterized by the author as a natural genus not merely by reason of characteristic external appearance but by reason of the structure of the capsule. The authors of Bryologia Germanica (1823) emphasize its affinity with *Weisia viridula*, and they include besides *H. rutilans* (Hedw.), also *H. squarrosum* Bryol. Germ., *H. obliquum* N. v. E.. *H. brachycarpon* Bryol. Germ., *H. subglobosum* Bryol. Germ., and *H. crispatum* Bryol. Germ.; however the last species is withdrawn again in 1831 by the authors in Bryol. Germ., I¹. 2, p. 33. Huebener in 1833 considers *H. obliquum* and *H. brachycarpon* as varieties, recognizes *Phascum rostellatum* Brid. as a *Hymenostomum*, designates it however as *H. microstomum* var. *β. mutilatum* Heuben. In Bry. Eur., 1846, *Gymnostomum tortile* Schwaegr. is after Fuernroh's example described as a *Hymenostomum*, however *H. rutilans* (Hedw.) is made a synonym of *Weisia mucronulata* Bruch., and *H. subglobosum* Bryol. Germ. a synonym of *Weisia viridula*; further *H. rostellatum* is treated as an *Astomum*, and by error *H. crispatum* Bryol. Germ. is also removed. C. Mueller in his Synopsis (1849) treated *Hymenostomum* as a subgenus of *Weisia*; so also Juratzka (1882). Schimper in 1860 puts it as a subgenus. under *Gymnostomum*, yet *G. rostellatum* is counted in; in the second edition of the Synopsis (1876) it is again raised to the rank of a genus and *H. unguiculatum* Phil. is added: by gross error *H. crispatum* Bryol. Germ. and *H. murale* (Spruce) are placed here. In Milde, who followed Lindberg, the genus is withdrawn in 1869, and its species are thrown in the subgenus *Euweisia* C. Muell, emend. Lindberg, in 1879, considered *Hymenostomum*, including *Euweisia* C. Muell. and *Astomum (Systegium)* as a subgenus of *Mollia* Schrank emend., a genus which in respect to structure of capsule and the anatomy of its stem and of leaf costa comprises in its limits diverse elements. Since according to present views a valid taxonomic arrangement cannot be founded on the structure of vegetative organs merely, *Hymenostomum* is entitled to generic rank, for here in the so-called hymenium of the spore case we find a structure which has its analogue only in the "drumhead," (the epiphragm or diaphragm) of the Polytrichaceæ. Strictly speaking *H. microstomum* and *H. squarrosum* are the only representatives within our territory, for *H. rostellatum* may be viewed as an *Astomum* and *H. tortile* probably does not belong here at all. That *H unguiculatum* Phil. represents a *Trichostomum mutabile* with rudimentary peristome is known; yet *Hymenostomum Muelleri* Bruch from Corsica (Flora, 1829, t. 2

f. 1), judging from the scant material which I examined, I consider a good species. This view is also supported by the description of this species in DeNot. Epil. p. 606."

This historical review is of interest and indeed is essential to a right understanding of what I wish to say regarding a moss which has given me not a little trouble since I found it in April, 1901. It is the *Astomum* sp. mentioned in "Some Notes on Collecting" in the BRYOLOGIST **6**: March, 1903. In November last I found this same moss on the banks of the Mississippi, on Prairie Island, some three miles above Winona. This last collection included many with immature but well nourished and developed capsules, also quite a few fully ripe ones, full of spores. These ripe capsules showed clearly a tendency to a ring of modified cells separating the lid. In February 1902, I sent some of the plants collected on the bluff to Mr. Ernest Salmon, at the Kew Herbarium, to have a comparison made with *Phascum subexsertum* Hook., the type of which I had reason to expect at Kew. Shortly after Mr. Salmon wrote that *P. subexsertum* Hook. cannot now be found. Meanwhile the discovery of a separable lid on the fresh material found, makes it clear that the plant is not a *Phascum* nor an *Astomum* but a *Hymenostomum*. I make it with considerable confidence close to, if not identical with, *Hymenostomum rostellatum* (Brid) Schimp. I reached this conclusion after a painstaking comparison of the plant with Limpricht's description, Laubmoose, p. 224, 225, and with an Italian plant collected near Milan by Mr. F. A. Artaria. Size of spores, cells of exothecium, leaf areolation and shape, all agree perfectly. In both plants the lid separation is equally well indicated. Both lack however the crucial mark the *hymenium*, whether on account of immaturity or for whatever other reason I cannot tell.

Not satisfied that my Italian plants were typical *H. rostellatum* I sent the Minnesota plant to the New York Botanical Garden for comparison with typical European material if possible. Mrs. E. G. Britton very kindly compared the plants, not only favoring me with her judgment but she also sent me a plant or two of each, enough for comparison, of *Astomum multicapsulare*, *Systegium* (*Astomum*) *Ludovicianum* Sulliv., and No. 766 of Crypto. Gerv. des Fichtelgebirges, *Phascum rostellatum* Brid., collected by H. C. Funck. Mrs. Britton correctly points out that No. 766 has a somewhat longer beak than my plant but otherwise it is identical. I can however not agree that it is not near *Systegium Ludovicianum* which has the same leaves and spores of the same size exactly, but its beak is just a trifle shorter. The tendency to a separable lid is here also noticeable, and the conviction has grown on me that *Systegium Ludovicianum* and *Hymenostomum rostellatum* are practically identical, the beak being of somewhat variable length on the material of my own collections, approaching on the one hand the short beak on the single capsule I have seen of Drummond's southern plant (New Orleans, 1841) communicated as *Systegium Ludovicianum*, and on the other hand the somewhat longer beaked plants from the Fichtelgebirge in Europe, the Italian plants holding in this respect the same middle place as my plants. It should be stated that so far as I could measure the spores on the plants sent me by Mrs. Britton I found them to agree exactly in size and rough surface with my plant and that from Italy, but that instead of being 17-22μ as Limpricht records, p. 225 of Vol. I., they are uniformly a little larger, measuring 26-30μ.

What I have now stated will make clear why I think Hooker and Wilson were probably right when they referred our American plants to *H. rostella-tum* to which Lesquereux and James did not agree, retaining Sullivant's name (see the Manual, p. 52, under *A. Ludovicianum*). To be sure Limpricht in his historical sketch tells us that *Hymenostomum rostellatum* may also be conceived as an *Astomum* (referring to the European form) which illuminates Hooker and Wilson's name *Phascum crispum rostellatum*, but judging from his practice (See Vol. I., p. 224) he himself considers it a *Hymenostomum*. According to this view *Astomum* (or *Systegium*) *Ludovicianum* Sulliv. becomes a synonym of *Hymenostomum rostellatum* (Brid) Schimp., in addition to the list given by Limpricht. On the other hand if the plant is to be referred to the genus *Astomum* Hampe., or *Systegium* Schimp., where it may stand with perfect propriety, since here also the lid is "distinctly formed but not easily detached," it should be called: *Astomum rostellatum* Bry. Eur., or *Systegium rostellatum* Boulay in Muscinees del Est , p. 586 (1872).

That Hooker and Wilson referred the plant to *Phascum* (*Astomum*) *crispum* as a variety, is very suggestive of the close similarity of the vegetative organs, and a look at Limpricht's synonyms of the European form, ever since Bridel described it as a Phascum, is sufficient to show its ambiguous position. The frequently clustered capsules of the species are referred to, both by Limpricht and in Lesquereux and James, and the somewhat more vigorous and branching habit of Sullivant's plants is purely local, judging from the several American specimens examined. The difference in length of operculum, the European plant being on the whole a little longer beaked, is not emphasized either in plate 12, Bry. Eur., or in Sullivant's Icone, plate 12, a comparison of which shows how slight is this difference, so slight indeed that one would not care to base even a variety on it.

Winona, Minn.

(Taken from Minnesota Botanical Studies as Reprinted July 3, 1903.)

THE MOSS FLORA OF THE UPPER MINNESOTA RIVER BY JOHN M. HOLZINGER.

A. J. GROUT.

This is an interesting report based on collections made under the auspices of the Minnesota Botanical Survey and lists ninety-six species, of which nearly one-half are new to the State, including six new species, five of which are by Cardot and Theriot. Two of these new species are forms confessedly near Hypnum riparium, That the polymorphous riparium might be segregated into several fairly well defined groups is almost certain, but it hardly seems worth while, from a scientific point of view, to found new species on specimens from a single collection or two collections from neighboring localties even.

Following are the descriptions of the new species translated from the Latin. These descriptions of all the species are accompanied by good illustrations.

BRYUM MINNESOTENSE Card. & Ther.

Dioicous, densely cespitose; stems 5-10 mm. high, erect, radiculose, with numerous innovations; stem leaves erect-open when moist, appressed when dry, about 2 by 1-1.5 mm., ovate-lanceolate, short acuminate, entire or denticulate at apex, costa excurrent, margins revolute from base to apex; basal leaves larger, longer acuminate, somewhat twisted when dry; basal cells hyaline, elongated rectangular, median and upper short-hexagonal, strongly chlorophyllose, marginal linear in 4-5 rows forming a distinct border. Seta 2-2.5 mm. long, usually bent at base; capsule subhorizontal or cernuous, oblong-pyriform with a long plicate neck when dry; operculum conic; annulus broad, exostome as in *B. pendulum*, 0.36 mm. high, inner lamellæ anastomosing. Endostome adherent, cilia not appendiculate. Spores 19-20μ. Antheridial plants unknown.

Differs from *B. pendulum* by its narrower, longer capsule, provided with a longer neck, longer cilia and dioicous inflorescence.

At Granite Falls (June 13, 14, 1901).

BRYUM HOLZINGERI Card. & Ther.

Closely related to the preceding from which it differs in the synoicous inflorescence, leaves narrower at base with margin less longly revolute, plane above, acumen longer, more distinctly denticulate at apex, median cells about twice longer, margin broader, composed of 6-7 rows of cells. Capsule longer and narrower for its length.

The longer and narrower capsule distinguishes this species from *B. pendulum*.

At Cedar Lake (June 18), Hartford (June 27) Foster (June 29, 1901).

CATHARINAEA MACMILLANI Holz.

Dioicous, archegonial heads only found. Plants not branching, reaching 2 cm. in length; leaves involute and circinate when dry, erect-open when moist, margin bistratose and constructed of two rows of cells, serrate with paired teeth; lamellæ 7-10, 8-12 cells high, the terminal cell lightly papillose. Other characters not known.

This species is at once distinguished by its papillose leaves. It is dedicated to Prof. Conway MacMillan, director of the Minnesota Botanical Survey. On ground near Ortonville (June 25, 1901).

NOTE—The original spelling of the generic name is *Catharinaea*, not *Catharinea*.

FONTINALIS OBSCURA Card.

Plants rather soft, obscurely green to dark green, blankish below, stems 10-15 cm. high, flexuous, denuded at base, irregularly or pinnately branched, branches patent to patulous, obtuse or short cuspidate; leaves rather densely foliate, slighly concave, rather soft erecto-patent, imbricated at the apex of stems and branches, fragile, often ragged or broken off, stem leaves ovate to oblong-lanceolate, obtuse or short acuminate, entire 3-3 5 by 0.9-1.2 mm., branch leaves narrower, 2.1-3.5 by .6-.85 mm.; alar cells subquadrate or oblong, not strongly differentiated, others linear, subflexuous, upper shorter, all chlorophyllose. Other characters unknown.

Seems to be related to *F. Novae-Angliae* Sulliv., but readily distinguished from this species by the leaves (chiefly the branch leaves) which are narrower, longer acuminate, entire at apex, and by the much smaller and less distinct alar cells. It belongs to the section *Heterophyllae*.

In the Minnesota River Channel, at Granite Falls (July 12, 1901).

AMBLYSTEGIUM BRACHYPHYLLUM Card & Ther.

Related to *A. riparium* from which it differs in its shorter ovate-lanceolate leaves, 1-6-1.7 by 0.7 mm., short acuminate, apex obtuse or subobtuse; costa strong, broader at base, 50–80μ wide, extending 2/3 or 3/4 the length of leaf; median cells linear, 70–90μ. Fruit unknown.

By the blunt or subobtuse acumen this moss resembles *A. vacillans* Sulliv., but it has much shorter and broader leaves. From *A. brevipes* Card. & Ther., it is well distinguished by the larger size, the blunt acumen, the longer and narrower cells, and the stronger costa. The polymorphous *A. riparium* constitutes a vast group of forms, some of which are constant enough and are sufficiently characterized to be considered as secondary or tertiary species; such are *A. Kochii* B. & S., *A. vacillans* Sulliv., *A. brachyphyllum* and *A. brevipes* Card. & Ther., *A. floridanum* Ren. & Card., and probably *A. argillicola* Lindb.

Granite Falls (July 15, 1901),

AMBLYSTEGIUM BREVIPES Card. & Ther.

In the *A. riparium* section of the genus; stems slender, creeping, branches short, leaves erect-open, about 1.2 by 0.6 mm., broadly ovate, short-acuminate, entire; costa narrow, 30μ wide at base, vanishing beyond the middle, often extending 2/3-3/4 the length of the leaf; areolation lax, basal cells rectangular, some quadrate, median subhexagonal, 55-70 by 12-15μ, upper shorter and broader; perichaetial leaves broadly ovate, suddenly narrowed into a slender acumen, often irregularly denticulate at base of acumen; costa extends beyond the middle; seta short, capsule oblong-arcuate, constricted under the mouth when dry; operculum conic.

This species differs from small forms of *A. riparium* by the shortly acuminate leaves, the looser areolation, the shape of the perichaetial leaves, and the short pedicel. A species from the Caucasus, *A. argillicola* Lindb., of which we know only the description published by Dr. V. F. Brotherus in his valuable paper, "Enumeratio Muscorum Caucasi," seems to be nearer to this species but still stands distinct from it by its leaves which are minutely denticulate from almost the base, its longer costa vanishing below the apex, and its narrower perichaetial bracts, with a less distinct nerve.

Near Montevideo (June 15, 1901); Hartford (June 27, 1901).

LETT'S HEPATICS OF THE BRITISH ISLANDS.*

Marshall A. Howe.

*H. W. Lett. A list, with descriptive notes, of all the Hepatics hitherto found in the British Islands. Published by the author, Aghaderg Glebe, Loughbrickland; Co. Down. 1902. Pp. 1—VIII-1-90. Price, 7s., 6d.

This work, which is a descriptive manual rather than a "list," is an attempt to popularize the study of the Hepaticae of Great Britain and Ireland. The language of the work, as the author states in the preface, " is not that usually found in botanical books, and which needs the assistance of a dictionary of Botanical terms or a vocabulary, but plain simple English." The author's purpose to employ simple language is rather well carried out, though certain terms like "monoicous," "dioicous," and a few others that occur in nearly every description are probably as susceptible to impeachment in the *role* of "plain simple English" as would be several other telling technical terms that are studiously avoided. The use of the metric system for measurements, which deserves speedily to become " plain English," even though not generally considered so at present, is to be commended.

In general plan and typography, the book suggests Dixon and Jameson's popular Handbook of British Mosses, and like that, it is a work that will prove of service to the American as well as to the British beginner in the study of bryophytes. But unhappily for the less experienced student, for whom the work is primarily intended, there are frequent errors or inaccuracies of statement which may mislead rather than help. Some of those have been listed by Mr. Symers M. Macvicar in a review, published in the *Journal of Botany* for December, 1902. Among those not noted by Mr. Macvicar may be mentioned the following remark under *Anthoceros punctatus:* "When without fruit they might be taken for young states of *Pellia*, but may be distinguished with aid of a lens by the absence of true stomata-pores in their surface and by the large cells." As a matter of fact, the existence of "true stomata-pores " or pores of any other kind in *Pellia* has, so far as we know, never before been hinted at by any one, while in the epidermis of the Anthocerotaceae there are actually inconspicuous clefts (especially on the ventral surface) which authors have described as "stomata" or " mucilage-slits." A very important and easily applied test which always serves to distinguish a sterile *Anthoceros* from a *Pellia* or any other hepatic in the narrower sense, is of course the presence of a *single* large chlorophyll body in each epidermal cell, while in *Pellia* each surface cell contains *several or many* very much smaller chlorophyll-bodies. A further remark of Canon Lett's, under *Anthoceros Stableri*, is that " The species of Anthoceros, if dried, are almost impossible to distinguish from each other." We cannot say how it may be with *A. Stableri*, but we have found in practice that *Anthoceros punctatus* and *A. laevis*, the other two species of his list, are best distinguished by decided differences in the nature of the surface-markings of the spores and by their color, characters that can be determined as well from dried material as from the living.

The arrangement of the genera in some parts of the book evidently follows natural relationships, but the sandwiching in of *Anthoceros* between *Sphaerocarpus* and *Marchantia* is even more violent and unnatural than was the now long obsolete placing of the Gymnosperms between the Monocotyledons and the Dicotyledons. The nomenclature of the work is, in general, that which is most familiar to English-speaking students of the Hepaticae, but in supporting the efforts of some continental writers to revive Dumortier's *Madotheca* in place of the Linnaean *Porella*, the author is departing not only from the English, American, and Scandinavian usage of many years' standing, but also, we believe, from the usage which must eventually prevail elsewhere. The book contains descriptions of several species, the discovery of which in the British Isles was too late for their inclusion in Pearson's admirable monograph. The work, as a whole, in spite of some patent defects, we think will prove useful to the less technically inclined students of the Hepaticae both in the British Isles and in North America.   New York Botanical Garden.

PAPILLARIA NIGRESCENS (Sw.) JAEG. & SAUERB.
Meteorium Nigrescens Mitt.

ELIZABETH G BRITTON.

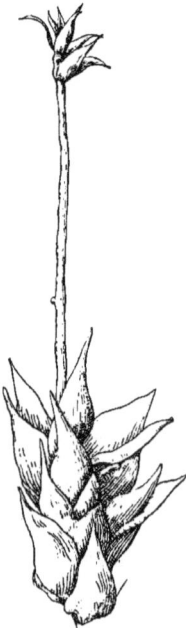

The type of this species was collected and described by Swartz from high mountains in Jamaica and was figured by both Hedwig and Schwaegrichen in the Species Muscorum. Various specimens from Mexico, the West Indies, Venezuela, Florida, and Louisiana have been also called by the same name. Since my note on "West Indian Mosses in Florida" was published (BRYOLOGIST, 6:4, 1903) it has become evident that the description of this species in the Manual is incomplete and misleading and that *Leptodon trichomitrion* has been and is liable to be mistaken for it. In fact the leaves are very much alike in shape and size, but the cells are different and those of *Meteorium* are papillose on both sides with three or four minute papillæ on each cell. Add to this the fact that the description of the fruit in the Manual is taken from Schwaegrichen's plate, as there are no fruiting specimens of this species in any herbarium that we have examined, and that all the species of this genus very rarely fruit, most of them being known only from sterile specimens, and we have an added reason for doubting the correctness of referring specimens from Lake Huron to this species. It seems likely that the hairy calyptra in *Leptodon trichomitrion*, has been the misleading character for confusing it with the description in the Manual.

The type of *Hypnum nigrescens*, Sw., is at Stockholm. It is described as being a foot long, with branches an inch long. None of the Florida specimens attain this size, and there is another character which is conspicuous in the Florida specimens which is not mentioned in the original description. These are the denuded flagellate branches shown in the accompanying illustration. The leaves fall off, leaving only a tuft at the apex, and this also falls, serving to propagate the species. On account of this character and also because of the more acuminate leaves, Austin distributed No. 533 of his Musci Appalachiani as *M. nigrescens* var. *Donnellii*. He had specimens for comparison from T. P. James and Wolle collected in Jamaica, Mexico and Venezuela. Although he did not publish any description of this variety, the notes in his herbarium show that he thought it sufficiently distinct at one time to be of specific rank. I have recently compared the Florida specimens with ones from Jamaica and find that ours have a smaller, narrower leaf, more subulate-acuminate with a narrower base, less auriculate angles which are decurrent with quadrate cells, and the cells of the auricles and apex are longer, less rhomboidal, with more numerous and prominent papillæ.

The best description of *M. nigrescens* is given by C. Müller in the Synopsis Muscorum. Kindberg has described two species from Florida, as he calls the variety *Donnellii* a sub-species, and refers them both to *Papillaria*. He states that *P. Donnellii* is not distinctly papillose (in this he is mistaken), and describes the flagellate branches in *P. nigrescens*. He is quite right in calling them Papillaria, but I do not believe there are two species in our Southern States and the characters he uses to differentiate the two species may be found in all the specimens I have seen. In fact Müller describes " slender flagellate stolons" in other tropical specimens. As we have very little material except from Florida in our collection, I feel some hesitation in deciding whether the differences are sufficient for even varietal rank, but only comparison with the type will decide this satisfactorily.

New York Botanical Garden.

A CORRECTION.

In the November number of THE BRYOLOGIST I wrote *Ulota coarctata* for *Ulota Ludwigii* Brid. This was a pure error as *coarctata* belongs to a different species if used with the generic name *Ulota*.

A. J. GROUT.

A CORRECTION.

Our attention has been called to an error in statement on page 101, in the November, 1903, number; sixth line from the top should read: (8) Leskea tectorum (A. Braun) Lindb.

A. M. S.

ANNUAL REPORTS OF THE SULLIVANT MOSS CHAPTER.

PRESIDENT'S REPORT.

It is with pleasure that I record the completion of one year of service to the Sullivant Moss Chapter. My duties have doubtless not been any more varied than were those of my predecessors, and I am persuaded that my somewhat critical attitude in regard to mosses sent me for determination has latterly rendered those duties somewhat less arduous than before. This I trust will be appreciated by all the members as a step in the line of progress.

I was about to say this and several other things, and bow myself off the pedestal of presidential honor and activity, when our charming secretary announced officially my re-appointment. My previous protests to those in whose hands I was had not availed. It is too late now, and the only way left open to me is to cheerfully and joyously be at the service of the members for another year. I stand ready to serve the Chapter faithfully according to my ability and time. JOHN M. HOLZINGER.

REPORT OF THE SECRETARY.

The membership of the Sullivant Moss Chapter has been strengthened during the year 1903 by the addition to its ranks of eighteen new members; fifteen have withdrawn, and one member, Miss Edith Barnes, of Northboro, Mass., has died. One hundred and twenty names are now enrolled.

The fact that almost constant requisition has been made upon the services of our expert bryologists in the determination of difficult questions, and that they have generously responded to demands upon their time and skill, is proof of a commendable degree of interest in the work of the Chapter. Interest has been further indicated by the number and value of the specimens which have been collected in widely separated stations and offered for distribution, and also in the readiness of members in applying for them. Ninety-seven species of mosses, lichens and hepatics have been distributed during the year, The moss herbarium contains about three hundred and forty-five species and varieties, representing one hundred and three genera; more than seventy new species have been added this year.

While the Chapter is under obligations to many others for contribution of specimens, it is especially indebted to Mr. A. J. Hill, of New Westminster, B. C., for rare and beautiful ones contributed to the several herbaria and , for general distribution. The secretary, in retiring from her office, acknowledges with much appreciation, the helpful services of those to whom she has looked for assistance, and she will continue to hold in remembrance the pleasant friendly relations which have grown up between her and the many members with whom she has been in correspondence during the past two years. HARRIET WHEELER,
Secretary.

REPORT OF JUDGE OF ELECTIONS.

Miss Harriet Wheeler,   December 1st, 1903.
        Secretary Sullivant Moss Chapter.

The following report of the election of officers of the Chapter for the year 1904 is respectfully submitted:

Whole number of votes cast ....14.
For President.—Prof. J. M. Holzinger ....14.
For Vice-Pres.—Mrs. C. W. Harris ....14.
For Secretary.—Miss Miller ....13.
  "   "   Miss Wheeler .... 1.
For Treasurer.—Mrs. Smith 13
  "   "   Miss Wheeler .... 1.

Prof. Holzinger, Mrs. Harris, Miss Miller and Mrs. Smith are elected.

                                        Edith A. Warner,
                                        Judge of Elections.

REPORT OF THE TREASURER.

The following statement for the year beginning December 1, 1902, and ending December 1, 1903, is respectfully submitted:

RECEIPTS.

By cash in hand December 1, 1902 .... $ 13.55
By dues from members 125 45
                                                      ———
                                                      $139.00

DISBURSEMENTS.

To Bryologist $101.70
To postage 7.80
To express 1.15
To Herbarium supplies 88
To stationery 1.90
To printing annual report for 1901 ....   5.00
  "   "   "   "   " 1902 ....   5.00
                                                      ———
                                                      $123.43
Cash in hand December 1, 1903 .... 15.57
                                                      ———
                                                      $139.00
                                    Harriet Wheeler, Treasurer.

REPORT OF THE LICHEN DEPARTMENT.

It is with pleasure that I can report continued interest in the work on the Lichens.

A year ago the Herbarium contained one hundred and seventy specimens, representing twenty-seven genera and ninety-five species and varieties. At the present time there are two hundred and sixty specimens, representing thirty genera and one hundred and ten species and varieties.

Judging from the many specimens sent me for determination and the letters received asking for assistance the interest is certainly increasing in the study of the Lichens.

Several members have offered Lichens for distribution, and it is hoped that many more will do so during the coming year.

Additions have been made to the Herbarium of a number of interesting specimens from British Columbia, and a set of sixty correctly determined specimens of the genus Cladonia, as well as many others.

As suggested in the article on Peltigera, if those interested in the Lichens will systematically study specimens of one particular genus over a period of several months, recording their observations, it will not only be of great assistance to themselves but will be useful to others. I shall be glad to hear from any one regarding his studies. Several have already sent interesting notes which will in time be compiled and published in THE BRYOLOGIST for the benefit of the lichen students.

Respectfully submitted,

CAROLYN W. HARRIS.

REPORT OF THE HEPATIC DEPARTMENT.

During the past year, through the efforts of various members of the Chapter, a goodly number of specimens have been added to the herbarium It now contains, in round numbers, 300 specimens, representing about 250 species. A large part of these are from foreign sources, and much of the material has been examined by the best hepaticologists of the day.

It is earnestly hoped that during the coming year, it can be brought up to at least 500 specimens, and to this end the cordial coöperation of all members is urged. It is also suggested that some plan be arranged by which the Chapter herbarium may be made directly available to the members.

During the past year the writer has, of necessity, been slow in answering inquiries concerning hepatics, but hopes to be able to make prompt replies in this matter hereafter.

Let every member make an effort to collect *some* hepatics for the Chapter herbarium; and, if possible, for the distribution. Even common species are desired.

Respectfully submitted,

WILLIAM C. BARBOUR.

LIST OF MEMBERS.

Adams, Miss Carrie E.................................Hinsdale, N. H.
Adams, Mr. F. M ....361 Madison street, Brooklyn, N. Y.
Ainslie, Mr. Charles N.............First National Bank, Rochester, Minn.
Ames, Mr. Oakes ....Ames Botanical Library, N. Easton, Mass.
Anderson, Mr. John A..... ....High School, Dubuque, Iowa.
Anthony, Mrs. Emilia C ..../............Gouverneur, N. Y.

Badè, Dr. Wm. F.......... ....Univ. of California, Berkeley. Cal.
Bailey, Dr. J. W ....The Arcade, Seattle, Wash.
Bailey, Miss H. B....................830 Amsterdam avenue, N. Y. City.
Barbour, Mr. Wm. C........ ....Sayre, Pa.
Barnes, Prof. Charles R.. ...Dept. Botany, Univ. of Chicago, Chicago, Ill.
Best, Dr. George N .... Rosemont, N. J.
Billings, Miss Elizabeth.... ....Woodstock, Ver.
Bonser, Mr. Thomas A.. ....Carey, Wyandot Co., Ohio.
Britton, Mrs. Elizabeth G.......Botanical Garden, Bronx Park, N. Y. City.
Brown, Mr. Edgar ......Div. of Botany, Dept. Agric., Washington, D. C.
Bruce, Mr. C. S..... ... .Shelburne, Nova Scotia.
Bryant, Miss E. B ... ...: ....32 Reedsdale street, Allston, Mass.
Carr, Miss C. M ....:........South Sudbury, Mass.
Carter, Mrs. R. H ....37 Church street, Laconia, N. H.
Chamberlain, Mr. Edw. B .... Cumberland Center, Me.
Chapin, Mrs. L. N........... ....11 East 32d street, N. Y. City.
Chase, Mrs. Agnes........ ... ...59 Florida Avenue, N. W., Washington, D. C.
Chase, Mr. Virginius H....................Wady Petra, Stark Co., Ill.
Cheney, Prof. L. S ....Barron, Barron Co., Wis
Choate, Miss Alice D......... ....3,400 Morgan street, St. Louis, Mo.
Clapp, Mrs. J. B. ....52 Hartford street, Dorchester, Mass.
Clarke, Miss Cora H........... ....91 Mt. Vernon street, Boston, Mass.
Clark, Mr. H. S.... ....31 Wells street, Hartford, Conn.
Clarke, Mrs. Sarah L ....1 West 81st street, N. Y. City.
Collins, Mr. J. Franklin... ....468 Hope street, Providence, R. I.
Coomes, Mrs. Laura M....................Queens, Queens Co., N. Y. City.
Craig, Mr. T.... .. ..1019 Sherbrooke, Montreal, Can.
Cresson, Mr. Ezra T., Jr....... ....Box 248, Philadelphia, Pa.
Crockett, Miss Alice L............. .... Camden, Maine.
Cummings, Miss Clara E.............Wellesley College, Wellesley, Mass.
Curtis, Mrs. Elizabeth B.....................Box 47, Hendersonville, N. C.
Cushman, Miss Mary H..............300 North Fifth street, Reading, Pa.
Dacy, Miss Alice E... .... 28 Ward street, South Boston, Mass.
Demetrio, Rev. Charles H ....Emma, Mo.
Doran, Miss Genevieve...........13 Washington Avenue, Waltham, Mass.
Dupret, Mr. H ....Seminary of Philosophy, Montreal, Can.
Eaton, Mr. Alvah H................................. ... .Seabrook, N. H.
Eby, Mrs. Amelia F. ... . 141 North Duke street, Lancaster, Pa.
Edwards, Prof. Arthur M..............423 Fourth Avenue, Newark, N. J.
Evans, Dr. Alex. W.......... .....2 Hillhouse Avenue, New Haven, Conn.
Fink, Prof. Bruce ....Grinnell, Iowa.
Fletcher, Mr. S. W.......................................Pepperell, Mass.
Garver, Mr. H. B ....:.... .... Middletown, Pa.
Gerritson, Mr. Walter.......... ....66 Robbins street, Waltham, Mass.
Gilbert, Mr. B. D.... ....Clayville, N. Y.
Gilman, Mr. Charles W....................Palisades, Rockland Co., N. Y.
Gilson, Miss Helen S ....50 Williams street, Rutland, Vt.
Graves, Mr. James A....... ....Susquehanna, Pa.
Gregory, Mrs. H. T......................... .. ....Southern Pines, N. C.
Greenalch, Mr. Wallace.......... ...34 North Pine Avenue, Albany, N. Y.
Greever, Mr. C. O ....1,345 East Ninth street, Des Moines, Iowa.
Grout, Dr. A. J.......................360 Lenox Road, Brooklyn, N. Y.
Hadley, Mrs. Sarah B.............. ....South Canterbury, Conn.
Harris, Mrs. Carolyn W.......... ...125 St. Marks Avenue, Brooklyn, N. Y.
Harris, Mr. Wilson P....................48 Laurel street, Buffalo, N. Y.
Haughwout, Miss Mary R..............Wilson College, Chambersburg, Pa.
Haynes, Miss Caroline C.... ....16 East 36th street, N. Y. City.

Hill, Mr. E. J ....7,100 Eggleston Avenue, Chicago, Ill.
Hill, Mr. Albert J ....New Westminster, B. C.
Holzinger. Prof. J. M...............................Winona, Minn.
Horton, Mrs. Frances B...................13 Brook street, Brattleboro. Vt.
House, Mr. Homer D...........Botanical Garden, Bronx Park, N. Y. City.
Huntington, Mr. J. Warren.......................Amesbury, Mass.
Jackson, Mr. Joseph ....15 Woodland street, Worcester, Mass.
Joline, Mrs. A. H ....1 West 72d street, N. Y. City.
Jump, Mrs. Harvey D...............................Sayre, Pa.
Kennedy, Dr. George G.......................Readville, Mass.
Krout, Prof. A. F. K ....Glenolden, Delaware Co., Pa.
Lamprey, Mrs. E. S....................2 Guild street, Concord, N. H.
Lippincott, Mr. Charles D.......................Swedesboro, N. J.
Lorenz, Miss Annie....................96 Garden street, Hartford, Conn.
Lowe, Mrs. Josephine D ....Noroton, Fairfield Co., Conn.
Marshall, Miss M. A ....Still River, Mass.
Martens, Mr. J. W., Jr................Shrub Oak, Westchester Co., N. Y.
Mathews, Miss Caroline....................Waterville, Me.
Maxon, Mr. Wm. R........U. S. National Museum, Washington, D. C.
McConnell, Mrs. S. D..................781 Madison Avenue, N. Y. City.
McDonald, Mr. Frank E................417 California Avenue, Peoria, Ill.
McBride, Mr. James..............................Tewksbury. Mass.
Merrill. Mr. G. K.................564 Main street, Rockland, Maine.
Metcalf, Mrs. R. E...............................Hinsdale, N. H.
Miller, Miss Bertha S...............4 Inwood Place, Upper Montclair, N. J.
Miller, Miss Mary F ....1,109 M street, N. W. Washington, D. C.
Mirick, Miss Nellie...........28 East Walnut street, Oneida, N. Y.
Murray, Miss Elsie...............................Athens, Pa.
Nelson, Mr. N. L. T.................3,968 Laclede Avenue, St. Louis, Mo.
Newman, Rev. S. M......cor. 10th and G streets, N. W., Washington, D. C.
O'Connor, Dr. Helen Cox ....Garden City, N. Y.
Palmer. Mrs. Rebecca L ....615 Putnam Avenue, Brooklyn, N. Y.
Perrine, Miss Lura L ... State Normal School. Valley City, N. Dakota.
Plitt, Mr. Charles L.........1,706 Hanover street, Baltimore, Md.
Pollard, Mr. Charles L........U. S. National Museum, Washington, D. C.
Puffer, Mrs. James J....................Box 39, Sudbury, Mass.
Rapp, Mr. Severin ....Sanford, Orange Co., Fla.
Read, Mrs. R. M ....175 Tremont street, Boston, Mass.
Robinson, Mr. C. B......................Pictou, Nova Scotia.
Rondthaler, Miss E. W.............Moravian Seminary, Bethlehem. Pa.
Sanborn. Miss Sarah F.................54 Center street, Concord, N. H.
Schumacher, Miss Rosalie...........................Millington, N. J.
Seely, Mrs. J. A ....89 Caroline street. Ogdensburg, N. Y.
Smith, Mrs. Annie Morrill..............78 Orange street, Brooklyn, N. Y.
Smith, Mrs. Charles L...............286 Marlborough street, Boston, Mass.
Stevens, Mrs. M. L.................39 Columbia street, Brookline, Mass.
Stockberger, Prof. W. W........Bureau Plant Industry, Washington, D. C.
Streeter, Mrs. Milford B...............113 Hooper street. Brooklyn, N. Y.
Talbott, Mrs. Laura Osborne..........."The Lenox," Washington, D. C.
Thompson, Miss Esther H...................Box 407, Litchfield, Conn.
Thompson, Mrs. H. G......950 West Washington street, Los Angeles, Cal.
Van der Eike. Mr. Paul...............................Osceola, Wis.
Warner, Miss Edith A ....19 Schermerhorn street. Brooklyn, N. Y.
Wheeler, Miss Harriet..................Chatham, Columbia Co., N. Y.
Wheeler, Miss Jane....................248 Lake street, Albany. N. Y.
Williams, Mrs. M. E...............1,536 Pine street, Philadelphia, Pa.
Williams. Mr. R. S............Botanical Garden, Bronx Park, N. Y. City.

CHAPTER NOTES CONTINUED.

OFFERINGS.

To Chapter Members only—for postage.]

Prof. J. Franklin Collins, 468 Hope street, Providence, R. I. *Distichum capillaceum* B. &. S. Collected in Maine.

Mrs. Agnes Chase, 59 Florida Ave., N. W., Washington, D. C. *Asterella hemispherica* Beauv.; *Entodon repens* (Brid.) Grout, c.fr.; *Thelia asprella* (Schwaegr.) Sulliv.; *Entodon seductrix* (Hedw.) C. M., c.fr.; *Entodon cladorrhizans* (Hedw.) C. M., c.fr.; *Physcomitrium turbinatum* Brid., c.fr. Collected in Illinois by Mr. V. H. Chase.

Mr. Severin Rapp. Sanford. Florida. *Rhizogonium spiniforme* Bruch., c.fr.; *Bryum Sawyeri* R. & C., c.fr. Collected in Florida.

Mrs. J. D. Lowe. 200 A street, S. E., Washington, D, C. *Brynhia Nova-Angliae* (S. & L.) Grout, c.fr.; *Thuidium delicatulum* B. & S., c.fr. Collected in Noroton, Conn.

Miss Alice L. Crockett, Camden, Maine. *Dicranum scoparium* Hedw., c.fr.; *D. undulatum* Ehrb., c.fr. Collected in Camden, Maine.

Mrs. Carolyn W. Harris, 125 St. Mark's Avenue, Brooklyn, N. Y. *Peltigera aphthosa* (L.) Hoffm. ; *P. canina* (L.) Hoffm. Collected, Chilson Lake, Essex Co., N. Y.

NOTICE.

M. Bescherelle, whose death recently was announced in this journal, was interrupted in the preparation of an important bryological work, a "Sylloge" of all the species of mosses described by him. M. Cardot, to whom its completion was entrusted, states that it will contain 450 to 500 pages. and that it will need to be published by *subscription*. It will be possible to print the work at $5.00 a copy, provided that at least *fifty* of the minimum of 140 subscribers necessary to begin the printing can be found in the United States. I desire to announce that I will head this list, and will also receive names of other subscribers, at Winona, Minn.

John M. Holzinger.

WANTED

Copies of The Bryologist, Vol. IV, No. 1. Will anyone having extra copies of The Bryologist for January, 1901, please communicate with the publisher. Copies of any other issue will be given in exchange or cash paid for the same. Mrs. Annie Morrill Smith,
78 Orange street, Brooklyn, N. Y.

VOLUME VII.   NUMBER 2

MARCH , 1904

THE BRYOLOGIST

AN ILLUSTRATED BIMONTHLY DEVOTED TO

NORTH AMERICAN MOSSES

HEPATICS AND LICHENS

EDITORS:

ABEL JOEL GROUT and ANNIE MORRILL SMITH

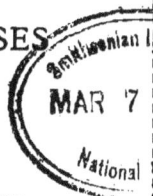

CONTENTS

Entered at the Post Office at Brooklyn, N. Y., April 2, 1900, as second class of mail
matter, under Act of March 3, 1879.

Published by the Editors, 78 Orange St., Brooklyn, N. Y., U. S. A.

PRESS OF MC BRIDE & STERN, 97-99 CLIFF STREET, NEW YORK

THE BRYOLOGIST

BIMONTHLY JOURNAL DEVOTED TO THE STUDY OF NORTH AMERICAN MOSSES
HEPATICS AND LICHENS.

ALSO OFFICIAL ORGAN
OF THE SULLIVANT MOSS CHAPTER OF THE AGASSIZ ASSOCIATION.

Subscription Price, $1.00 a year. 20c. a copy. Four issues 1898, 35c. Four issues 1899, 35c. Together, eight issues, 50c. Four issues 1900, 50c. Four issues 1901, 50c. Four Vols. $1.50 Six issues 1902, $1.00. Six issues 1903, $1.00.

Short articles and notes on mosses solicited from all students of the mosses. Address manuscript to A. J. Grout, Boys' High School, Brooklyn, N. Y. Address all inquiries and subscriptions to Mrs. Annie Morrill Smith, 78 Orange Street, Brooklyn, N. Y. For advertising space address Mrs. Smith. Check, except N. Y. City, MUST contain 10 cents extra for Clearing House charges.

Copyrighted 1904, by Annie Morrill Smith.

THE SULLIVANT MOSS CHAPTER.

President, Prof. J. M. Holzinger, Winona, Minn. Vice-President, Mrs. C. W. Harris, 125 St. Marks Avenue, Brooklyn, N. Y. Secretary, Miss Mary F. Miller. 1109 M Street, Washington, D. C. Treasurer, Mrs. Smith, 78 Orange Street, Brooklyn, N. Y.
Dues $1.10 a year, this includes a subscription to THE BRYOLOGIST.
All interested in the study of Mosses, Hepatics, and Lichens by correspondence are invited to join. Send dues direct to the Treasurer. For further information address the Secretary.

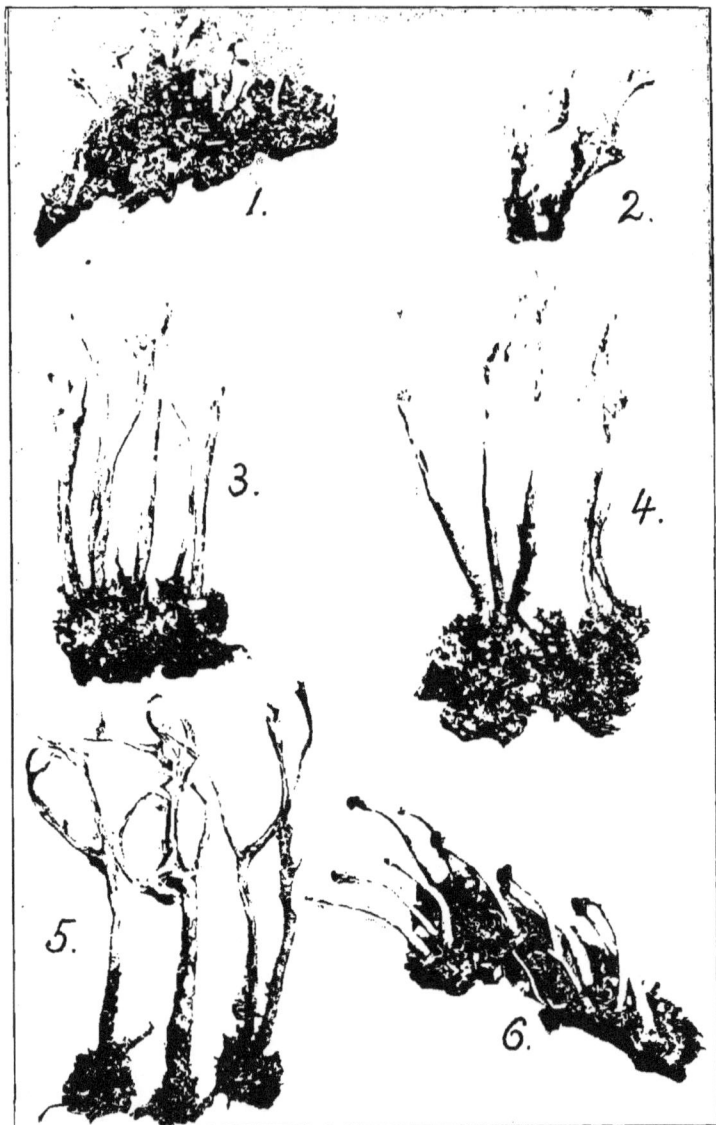

PLATE III. Fig. 1. *Cladonia fimbriata* var. *simplex*. Fig. 2. Var. *prolifera*. Fig. 3. Var. *cornutoradiata*. Fig. 4 Var. *radiata*. Fig. 5. Var. *subulata*. Fig. 6. Var. *coniocraca*. Enlarged ⅓.

FURTHER NOTES ON CLADONIAS.
˙Cladonia fimbriata.

BRUCE FINK.

It is the intention in the present paper to follow out, with reference to a single species, the work begun in a previous number of the BRYOLOGIST, (6:2.1903). With all due respect to the labors of the noted American lichenist, Tuckerman, it must be apparent to all who have attempted to use his diagnoses of American *Cladonias* as aids in determination, that they are too brief and indefinite. Tuckerman recognizes in his manual just two varieties of *Cladonia fimbriata*, disposes of the species in a half page, and gives not the slightest hint that the forms are extremely varied and difficult to determine. This view is all that could be expected from one who was a pioneer in the study of American lichens, and much as Tuckerman has done for American lichenology, we can not afford to do otherwise than pass beyond his results as rapidly as may be with some adequate degree of certainty.

In passing beyond the Tuckermanian view, we have been so fortunate as to have the aid of Dr. E. Wainio, and we now have his view of more than two hundred specimens of American *Cladonias*, which the writer has sent to him from time to time. Attention was directed to the extremely great amount of variation in forms of *Cladonia fimbriata* years ago in work in the field, and an especial effort was made to obtain all of the forms possible. But it was only by a careful study of the species, as viewed by Dr. Wainio, and set forth in great detail in one hundred and three pages of his monograph of the genus Cladonia, that the present writer began to realize something of the difficulties to be encountered in the attempt to gain anything like an adequate knowledge of the species. In Wainio's monograph, sixteen varieties and a very large number of subvarieties and forms are recognized. We have not been able to see the subvarietal distinctions in some instances even with specimens which have passed through Dr. Wainio's hands before us, and so it is not deemed wise to burden these pages with them. However, though we may not be able to follow the specialist in the genus into all of the intricacies of the most minute and discriminating observations, we can at least improve matters somewhat, and perhaps as much as is desirable, by attempting somewhat brief and yet sufficiently definite descriptions of the twelve varieties which are well known to exist in North America.

By giving figures of our more common forms with the descriptions, it is hoped that the student of lichens will not confuse the varieties and assign them to other species so frequently as has been done in the past. The figures are not in this instance all from plants examined by Dr. Wainio, two or three of them having been selected from other specimens which seem to bring out the varietal characters better. Some of the figures are from European plants, these having been selected because better specimens from which to secure the photographs.

Concerning some of the twelve varieties recorded below, they are either rare in North America or little is known of their distribution. A large amount of material in various herbaria the writer has not been able to see, and some of this would no doubt throw much light on the matter of distribution, especially that of Tuckerman's collection in New England. Passing to descriptions, the general description of the species will be given first, followed by shorter diagnoses of the varieties and such statements regarding distribution as can be made in the present state of knowledge of the species.

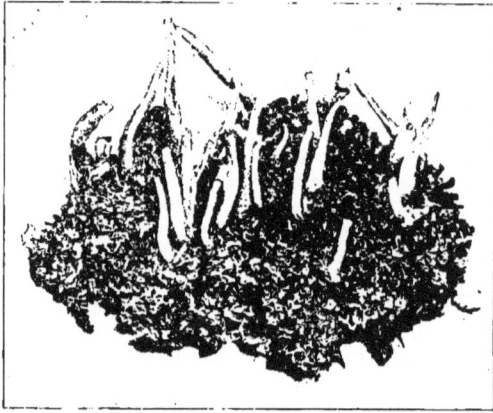

Fig. 7. *Cladonia fimbriata* × 2.

CLADONIA FIMBRIATA (L.) Fr. Lich. Eur. Ref. 222. 1831. Fig. 7.

Primary thallus commonly persistent, composed of digitately or irregularly incised or lobate, flat or concave, frequently involute or convolute, ascending, clustered or scattered, medium sized squamules, which are 2-9 mm. long and nearly or quite as wide, sea-green above or varying toward olivaceous or whitish, below whitish or darkening toward the base and the whole lower surface and edges sometimes sorediate-granulate. Podetia arising from the surface of the squamules, 4-100 mm. long and .5-3.5 mm. in diameter, cylindrical to tubaeform or rarely turbinate, the sides rarely rimose, commonly occurring in larger or smaller clusters, erect or rarely ascending or irregularly curved, commonly decorticate and more or less sorediate, or areolate or verrucose-corticate toward the base, or the basal corticate portions even subcontinuous, destitute of squamules or more or less squamulose, especially toward the base, sea-green varying toward whitish or brownish, the decorticate portions commonly whitish, sometimes cup-shaped, or the apices frequently cornute or subulate. Cups well developed or abortive, abruptly or gradually dilated, regular or irregular, the cavity commonly deep and non-perforate, the margin entire, dentate or proliferate, the proliferations one to several and the ranks one to three. Apothecia rather rare and medium sized, .8-2 mm. in diameter, solitary and rounded or irregularly conglomerate, sessile or pedicellate on the margins of the cups or at the cornute or subulate apices, flat and immarginate or more commonly becoming convex and immarginate, brown or rarely reddish-brown. Hypothecium pale or cloudy. Hymenium pale or pale-brownish below and brownish above. Paraphyses rarely branched, commonly thickened and brownish toward the apex. Asci clavate or cylindrico-clavate.

Generally distributed over North America and throughout the world in one form or another, the varieties being connected by various intermediate forms and altogether constituting perhaps the most confusing assemblage of lichens known to our flora.

CLADONIA FIMBRIATA (L.) Fr. var. SIMPLEX (Weis.) Wainio Mon. Clad. Univ. 2:356. 1894. Plate III. Fig. 1.
Podetia erect and straight, scarcely exceeding 3-30 mm. in length, scyphiform (i. e. cup-like), the cups well developed, 2-7 mm. in diameter, regular or becoming suboblique, with entire or dentate margin. Apothecia rare, sessile or pedicellate on the margins of the cups. Confusingly like forms of *Cladonia pyxidata*, but may usually be distinguished by the more slender habit, the more sorediate-granulate condition and transitional states passing into strictly cylindrical forms of the present species.

Found on various moist and somewhat shaded soils and more commonly on decaying wood. Examined by the writer from Newfoundland, Ontario, Minnesota, Iowa, Florida, Colorado and Idaho. Cited by Wainio from such widely separate localities as Great Bear Lake and New Mexico and California. These localities give the variety a general North American distribution. Known also in all of the grand divisions.

CLADONIA FIMBRIATA (L.) Fr. var. PROLIFERA (Retz.) Wainio Mon. Clad. Univ. 2:270. 1894. Plate III. Fig. 2.
Podetia 20-70 mm. long, scyphiform, repeatedly proliferate from well developed cups, commonly straight and erect, wholly decorticate and for most part sorediate, or having a minutely areolate or verrucose cortex below, sometimes squamulose especially toward the base. Cups 2-10 mm. in diameter, commonly somewhat abruptly dilated, regular or oblique, proliferations one or more from each cup and the ranks two or three or rarely more, the upper ranks usually quite as long as the lower and scyphiform, but the terminal cups commonly narrowed. Apothecia rare and usually borne on the cups of the highest rank. Readily distinguished from the last by the proliferous habit.

Occurring on damp earth and more rarely on mossy rocks and decaying trunks of trees. Seen by the writer from Newfoundland and from northern Minnesota. Also cited by Wainio from Vancouver Island. Supposed to be widely distributed in Europe, but Wainio gives only five stations. Known also in South America and Asia, but hardly a common lichen in any country as yet.

CLADONIA FIMBRIATA (L.) Fr. var. CORNUTORADIATA Coem. Clad. Ach. 40. 1865. Plate III. Fig. 3.

Podetia elongated (ours and European material seen 30-50 mm. long), sometimes bearing narrowed or abortive cups, simple or branched and the branches cornute or scyphiform, destitute of squamules or squamulose toward the base, decorticate and sorediate, or corticate toward the base and rarely also below the cups, the cavity of the cups also sorediate.

The specimens collected by the writer at Kettle Falls and at Tower, both in northern Minnesota,'grew on earth, and the same is true of the only European specimen seen by the writer. Not likely to occur on decaying wood. Not known elsewhere in North America, and in foreign countries only in Europe. Known to us through the kindness of Dr. Wainio, who determined one of the collections from Minnesota.

CLADONIA FIMBRIATA (L.) Fr. var. RADIATA (Schreb.) Wainio Mon. Clad. Univ. 2:277. 1894. Plate III. Fig. 4.

Podetia commonly elongated, 17-75 mm. in length, scyphiform, elongate-turbinate or subtubaeform, frequently more than one-ranked and the sterile apices cornute, subulate or rarely obsoletely scyphiform, commonly straight and suberect, wholly decorticate and sorediate, or corticate and minutely areolate or verrucose toward the base, without squamules or rarely squamulose, especially toward the base. Cups rather small, 2-5 mm. in diameter, gradually or quite abruptly dilated, regular or irregular, the margins dentate to proliferate, the proliferations one to several and elongated or quite short, often two or three ranked. Apothecia rare, sessile or rarely short pedicellate on the margins of the cups.

Usually occurring on earth or mossy rocks, but once collected by the writer on rotten wood. Examined from Minnesota and Newfoundland. Credited by Wainio from Kotzebue's Sound, from Canada, from the White Mountains and from California. These widely separate localities would seem to give the variety a wide distribution throughout the northern portion of the United States and in British America. Known in all of the grand divisions except South America.

CLADONIA FIMBRIATA (L.) Fr. var. SUBULATA (L.) Wainio Mon. Clad. Univ. 2:282. 1894. Plate III. Fig. 5.

Podetia much elongated, 30-100 mm. in length. almost always cupless, cylindrical, simple or variously branched, the sterile apices obtusely cornute or subulate, erect and straight, or flexuous especially toward the apex, wholly decorticate and sorediate, or areolate or subcontinuously corticate toward the base, without squamules or more or less squamulose, especially toward the base. Apothecia rare, at the apices of the podetia.

The plant occurs on earth, especially over rocks on thin soil. Said to occur rarely on rotting wood. The American material examined by the writer is all from the northern half of Minnesota, where the plant is quite frequently seen. Wainio cites the variety from Vancouver Island and from the White Mountains. Probably not infrequent from Minnesota eastward about the Great Lakes into the mountains of New England and northward in British America. Known in all of the grand divisions.

CLADONIA FIMBRIATA (L.) Fr. var. NEMOXYNA (Ach.) Wainio Mon. Clad. Univ. **2**:295. 1894.

Podetia commonly 25-90 mm. long, scyphiform or subscyphiform, two or three ranked, the sterile apices abortively scyphiform, cornute or subulate, suberect or more or less flexuous, wholly decorticate and sorediate, or the basal half (more or less) variously areolate or verrucose corticate as also at the base of the apothecia and the proliferations, without squamules, or squamulose toward the base and below the cups, or the whole podetium very rarely and sparsely squamulose. Cups small or abortive, 1-3.5 mm. in diameter, gradually or somewhat abruptly dilated, commonly becoming irregular, the margin dentate or proliferate, the proliferations one or more and short or quite elongated. Apothecia rare, and sessile or on pedicels on the margins of the cups. Dr. Wainio would refer all of our material to the subvariety, *fibula* (Ach.) Wainio Mon. Clad. Univ. **2**:300. 1894. In this the podetia are commonly simple and scarcely ever exceed 50 mm. in length, and the cups are more regular and scarcely ever proliferate.

On shaded earth, especially on thin soil over rocks in woods. Known to the writer only through his specimens from the northern half of Minnesota, where the variety is rather rare. Cited from New Bedford, Massachusetts, by Wainio. Apparently not a common variety anywhere, but still recorded from all of the grand divisions.

CLADONIA FIMBRIATA (L.) Fr. var. CONIOCRAEA (Flk.) Wainio Mon. Clad. Univ. **2**:308. 1894. Plate III. Fig. 6.

Podetia rather short, commonly 5-25 mm. long and 1-2 mm. diameter, cupless and cylindrical or abortively scyphiform, simple or rarely and sparsely short-branched toward the apex, the sterile apices subulate, cornute or abortively scyphiform, commonly straight and erect, but sometimes flexuous, wholly decorticate and sorediate, or corticate toward the base and rarely below the cups; the cortex subcontinuous or areolate-verrucose, without squamules or more or less squamulose, especially toward the base. Cups rare and small or abortive, 1-2 mm. in diameter, terminal with an entire or at least non-proliferous margin. Apothecia rather rare, at the apex of the podetia or on the margin of the cups, subsolitary and on very short pedicels.

Commonly on old and rotting wood or among mosses over rocks. The writer finds the plant generally distributed over Minnesota and Iowa, and has examined it from Newfoundland, New England, Ohio, Illinois and Colorado. Some material from California also seems quite as much at home here as in the next. This and the next include a large part of *C. fimbriata* var. *tubaeformis* of Tuckerman's manual, and no doubt both of the varieties occur to the south also and have a very general North American distribution. Known also in Europe, Asia and Australia.

CLADONIA FIMBRIATA (L.) Fr. var. APOLEPTA (Ach.) Wainio Mon. Clad. Univ. 2:307. 1894.

Podetia commonly quite short, cupless or narrowly or abortively scyphiform, wholly decorticate and sorediate, or corticate toward the base and rarely below the apothecia, which are rare and brown, brick-colored, or pale. The podetia rather shorter and more slender than the last, lighter in color and more frequently squamulose.

The last sentence above is based on six or seven specimens examined by Dr. Wainio, mostly from Minnesota. The habitat is as that of the last, and the writer freely admits that he can not distinguish between the two from any description at hand and only in the best marked specimens. It may well be doubted whether the two should be separated, and this view is strengthened by a perusal of Dr. Wainio's descriptions. Material which seems nearer the present form has been examined from Minnesota, Iowa, New England, Ohio, Illinois and California, and the North American and foreign distribution is doubtless about the same as that of the last.

CLADONIA FIMBRIATA (L.) Fr. var. OCHROCHLORA (Flk.) Wainio Mon. Clad. Univ. 2:319. 1894.

Podetia commonly rather short, about 5–40 mm. in length, cylindrical or tubaeform, cupless or scyphiform, rarely more or less rimose, sometimes more or less flexuous, frequently more or less squamulose, partly decorticate and sorediate and in part corticate, especially toward the base and below the apothecia, or some of the podetia wholly decorticate, the sterile apices cornute or subulate. Cups abortive or rarely well developed and dentate or proliferate, the proliferations sometimes numerous, ranks one to three, the lower longer; cavity of cups commonly sorediate. Apothecia rather rare, and of medium size, about .7–4 mm. in diameter, solitary or more less conglomerate at the ends of the podetia or sessile on the margin of the cups, brown, brick-colored or pale.

The plant occurs on old wood and among mosses over rocks. especially in more or less shaded places. Dr. F. Arnold listed this variety from several localities in Newfoundland, and Dr. Wainio credits it from Massachusetts, Washington and California. American specimens have not been seen by the writer. Known in all of the grand divisions.

CLADONIA FIMBRIATA (L.) Fr. var. BALFOURII (Cromb.) Wainio Mon. Clad. Univ. 2:339. 1894,

Podetia about 5–25 mm. long and .5–1.5 mm. in diameter, cupless and cylindrical, simple or rarely and sparcely branched toward the apex, erect or suberect, straight or rarely subflexuous, wholly decorticate or rarely corticate or subcorticate toward the base, the decorticate portions most commonly minutely and densely sorediate, without squamules or rarely more or less squamulose toward the base, more or less dull-waxy in appearance, especially when damp, the sterile apices narrowly subulate or obtusely cornute. Apothecia brownish or rarely pale.

Occuring on earth, among mosses over rocks or rarely on rotting wood. Cited from the White Mountains and Nicaragua by Wainio, but the writer knows the variety only through a specimen sent by Dr. Wainio and collected in Brazil. Known also in Africa and Australia.

CLADONIA FIMBRIATA (L.) Fr. var. BORBONICA (Del.) Wainio Mon. Clad. Univ. **2**:343. 1894.

Podetia about 5-30 mm. long and .5-1.5 mm. in diameter, subcylindrical,. cupless or narrowly scyphiform, simple or rarely and sparcely branched, straight or flexuous, commonly sterile, wholly decorticate or more or less corticate toward the base, where the corticate areas are then subcontinuous, verrucose or areolate, thickly squamulose especially toward the base, and the squamules lacerate and more or less sorediate, the podetia also similarly sorediate especially toward the apex, or the soredia disappearing, the corticate portions sometimes more or less waxy, the apices subulate or obtusely cornute. Cups about .8-2 mm. wide, sometimes abruptly dilated. subregular, margin entire or rarely dentate or proliferous. Apothecia dusky-brown.

Occurring on old trunks in more or less sunny places and on the naked sides of mountains. Cited from Mexico by Wainio, but the writer has not seen the plant. Also from South America, Australia, southern Asia and Africa.

CLADONIA FIMBRIATA (L.) Fr. var. CHLOROPHAEOIDES Wainio Mon. Clad. Univ. **2**:336. 1894.

Podetia about 5-50 mm. in length and .5-3 mm. in diameter, tubaeform or rarely subturbinate, scyphiform, suberect and .usually straight, wholly decorticate and minutely and densely sorediate, or the soredia finally disappearing wholly or in part, without squamules or minutely isidioid-squamulose toward the base, or even bearing larger squamules here, more or less distinctly waxy in appearance. Cups well developed, about 1-12 mm. in diameter, somewhat abruptly or rarely more gradually dilated, commonly regular, margin entire or rarely dentate, the cavity decorticate and sorediate. Apothecia borne on the margin of the cups, on pedicels which are 1-8 mm. long, solitary or radiately arranged, brown, varying toward reddish or yellowish.

Occurring on earth, rocks or rotting trunks of trees. Reported from Mexico and Nicaragua by Dr. Wainio. Not known to the writer. Also found in South America, Africa and Australia. Thus this variety and the last above seem to be southern forms, not likely to occur in North America, except in the most southern portions.

Aside from the collections of the writer, mainly in Iowa and Minnesota, the material which he has examined, or cited on authority of Dr. Wainio, was collected for the most part by W. G. Farlow, H. Willey and Clara E. Cummings in New England; A. C. Waghorne, in Newfoundland; J. Macoun, in Canada; C. F. Baker, in Colorado; A. A. Heller, in Idaho; E. E. Bogue, in Ohio; H. E. Hasse, in California, and G. P. Clinton. in Illinois. The plants photographed from Europe were collected by L. Scriba, of Höchst in Germany, whose plants are among the best that come to us and are largely determined or examined by Dr. Wainio.　　　　Grinnell, Iowa.

THE GENUS ANACOLIA IN NORTH AMERICA.

John M. Holzinger.

Among some plants recently sent me by Mr. E. P. Sheldon I found a moss which was evidently a Bartramia but which from its gross appearance I refused to refer to *Bartramia Menziezii* Turn. It has a comparatively scant supply of red radicles clothing the stems below the shoots; its leaves appear more rigid and darker green; and the capsules are markedly longer cylindrical. After repeated study, looking up all the available literature on Bartramia and related genera, I concluded that I must have *Glyphocarpa Baueri* Hampe. which is cited as a synonym of *Bartramia Menziezii* in L· & J. Manual, p. 204. The note in the Manual in fine print under the species but strengthens me in my supposition that I have rightly diagnosed this plant. (E. P. Sheldon's No. 10050).

The following characters are clearly established in the course of my examinations: the stem sections show the epidermis beset with papillae, and are octagonal, showing an eight-ranked leaf arrangement: the leaves are almost devoid of papillae, only a few occurring along the costa on the upper, inner surface, so that an ordinary observation would lead to the conclusion that there are none at all: lastly the plants are evidently dioicous. Now Limpricht in his diagnosis of the genera of Bartramiaceae assigns to Bartramia only plants with synoicous or autoicous inflorescence, with striped and furrowed capsules and with leaf base mostly sheathing. In none of these characters did the plant before me agree with Bartramia, but on careful comparison with Californian *B. Menziezii* I found to my surprise that it agreed in all these points and that furthermore its peristome is as figured in Sullivant's Icones, Suppl. t. 26. Indeed in dissecting one capsule I found one solitary ghost-like translucent real tooth, the mates of which must have been left behind in the lid, and which must be the "pellucid membrane" referred to in L. & J.'s footnote, only here the full set of so-called teeth as figured by Sullivant were also present. Therefore, I am led to express the opinion, which needs verification by other observers, that Sullivant's figure really shows segments, not teeth. This view is the more plausible since they look more like "segments" of certain other genera than like bryaceous "teeth."

A still closer comparison of the California and Oregon plants led to the discovery of at least two specific differences: the capsule in the California plant has about 1/4 of the entire length, namely that part which is below the loosely hung spore sac contracted into a distinct collum; and the spores measures 28-30μ; while the capsule of the Oregon plant shows no collum at all, the spore sac descending lower down and the remnant does not contract into a collum; and the spores measure only 18-22μ. For these reasons I hold that the Oregon plant must after all be distinct from *B. Menziezii* and can not well be produced simply by differences in exposure, as is suggested by Lesquereux and James.

Now I felt these plants could not stand under Bartramia, in Limpricht's sense, neither could I bring them under the next genus, Anacolia, unless I restricted myself to the characters predicated in the Key to Genera (Laub-

. moose, Vol. II, p. 534), "Capsule unstriped, not furrowed, dioicous." These three points agree perfectly with both our west coast plants. Nor was I encouraged to place them here on reading the original description of the genus in Sch. Syn. 2 Ed. p. 513: "Plants caespitose, quite stout, very radiculose. Leaf arrangement and leaf structure as in the genus Bartramia. Flowers dioicous, the male ones gemmiform. Capsule on a short pedicel (whence the Greek name meaning "short foot"), erect, spherical, symmetrical, not striped, when dry not sulcate, but strongly wrinkled or rugulose, leptodermous, with a much shorter spore sac. Peristome none. Spore as in Bartramia. The genus is very distinct from Glyphocarpa Rob. Brown. It differs from Bartramia by the capsule being quite leptodermous, unstriped, when dry not furrowed and by the mouth being always naked. Several exotic species are known." Thus says Schimper.

The only species the description of which is accessible to me is *Anacolia Webbii* (Mont.) Schimp. in Limpricht. Laubmoose, Vol. II, p. 547, has ascribed to it "very strongly prickly-papillose radicles, lamina mamillose on both sides," neither of which characters is found on our two American plants. From the genus as characterized by Schimper they seemed at first sight excluded by the presence of a peristome, and by the shape of the capsule, however fitting into it in other respects. So I find myself in a measure reconciled to the view indicated in Paris' Index which correctly gives, first *Anacolia Baueri* (Hpe.) Par., as the name for our Oregon plant; second, *Anacolia Menziezii* (Turn.) Par., as that for the more southerly plant.

The admission of these American plants into the genus Anacolia demands a modification of Schimper's characterization, somewhat as follows: **Anacolia Schimp.** emend—Plants caespitose, more or less strongly cohering below the shoots of the season by a felt of brown radicles which are either prickly-papillose or minutely roughened. Branching monopodial and dichasial. Stem eight-angled, rough-papillose. Leaf arrangement eight-ranked; leaf form and reticulation as in Bartramia, but leaf base not sheathing at all. Inflorescence dioicous, antheridial buds gemmiform. Capsule *usually* on a short pedicel (the names does not literally apply to the American representatives), erect, spherical or cylindrical, symmetrical or slightly curved, unstriped, when dry not furrowed but strongly rugulose, leptodermous, with a shorter, loose spore sac and a persistent columella (in the American species). Peristome none or consisting of 16 so-called teeth, and these inserted half their length below the mouth of the capsule.

In this sense we can admit our two west coast mosses into the genus Anacolia. Of course the question of the full specific value of *Anacolia Baueri* is unsettled; that it is different from *A. Menziezii* is certain, but whether to write it as a variety of the latter species becomes largely a matter of taste. Winona, Minn.

NOTES ON SOME NORTH AMERICAN MOSSES.

JULES CARDOT.

DICRANUM DEMETRII R. & C. (See BRYOLOGIST 6. 1903, p. 85.)

In the Bulletin de la Société Royale de Botanique de Belgique, Vol. 36, part 2, p. 173, we reduced this moss as a variety of *Cynodontium virens*.

DICRANUM SUBFULVUM R. & C. (See BRYOLOGIST 6, p. 87.)

The locality has been omitted in the Botanical Gazette. This moss was collected in 1894 by Rev. C. H. Demetrio, on sand rocks near Perryville, Perry Co., Missouri.

TRICHOSTOMUM INDIGENS R. &. C. (See BRYOLOGIST 6, p. 87).

We subsequently redescribed this moss as *Barbula indigens*, in Bull. de la Soc. Roy. de Bot. de Belgique, Vol. 36, part 2, p. 176, and it was distributed under this name in our Musci Americae Septentrionalis Exsiccati, No. 280.

PILOTRICHELLA CYMBIFOLIA (Sulliv.) R. & C. (See BRYOLOGIST 6, p. 60.)

I have this species from two localities in Florida, Enterprise (Fitzgerald) and Beauclerc (Sawyer), and from three localities in Louisiana, Baton-Rouge, Fontainebleau, (St. Tammany Co.,) and Chinchuba near Mandeville (Langlois). As yet the fructification is unknown.

METEORIUM NIGRESCENS (Sw.) Mitt. (See BRYOLOGIST 6, p. 60).

I entirely agree with Mrs. E. G. Britton when she thinks that this moss has never been gathered in Canada; but when she says that thus far the same species has not been collected by anyone else except in Florida, this is not exact: in our Catalogue, Musci Americae Septentrionalis, p. 45, we recorded this species from Louisiana, where it was collected at Home Place by the late Rev. Mr. Langlois, in 1884. I have it also from Beauclerc, Florida, leg. Sawyer. It is not a Meteorium, but a Papillaria (*P. nigrescens* Jaeg.).

METEORIUM PENDULUM Sulliv.

This species which is also a Papillaria (*P. pendula* R. & C.) has been indicated in western Louisiana, without special locality. I received from Rev. Mr. Langlois numerous fine specimens collected at Baton-Rouge, on *Arundinaria macrocarpa*. (Ren. & Card. Musci Amer. Sept. Exsicc. No. 79).

ANOMODON TOCCOAE Sulliv. & Lesq. (See BRYOLOGIST 5, 1902, p. 12.)

I have also good specimens of this moss collected at Baton-Rouge by Langlois.

LESKEA DENTICULATA Sulliv.

This moss is not at all a Leskea but a Schwetschkea (a genus of Fabroniaceae), and thus must be called *Schwetschkea denticulata* (Sulliv.) Card. *S. Japonica* Besch., of which I have carefully studied authentic specimens, is merely a synonym of the Sullivant species.

HOMALOTHECIUM SUBCAPILLATUM Sulliv.

In THE BRYOLOGIST, 1903, p. 65, Dr. A. J. Grout proposes for this species a new generic name: Burnettia. He has evidently forgotten that in 1899 I established for the same plant the section Homalotheciella (Bull. Herb. Bois-.

sier, Vol. 7, p. 374). Therefore, if this plant should be raised to the generic
rank, according to the rules of the Paris Code the name must be: *Homalotheciella*.

In the paper quoted above, I divided the genus Homalothecium into two
sections, characterized as follows:

EUHOMALOTHECIUM Card.

Mosses of large size; leaves not or hardly concave, generally deeply
plicate lengthwise; areolation linear; flowers dioicous; lid conic; *H. sericeum*
Br. Eur., *H. Philippeanum* Br. Eur., etc.

HOMALOTHECIELLA Card.

Mosses of small size; leaves concave, not plicate lengthwise; areolation
much looser; flowers monoicous; lid rostrate: *H. subcapillatum* Sulliv., and
perhaps *H. tenerrimum* (C. M.) Jaeg.   Charleville, France, Nov. 18, 1903.

NOTE.—My idea was not to propose a new generic name for *Homalothecium subcapillatum* alone but for the whole genus, as Homalothecium is
untenable.   As M. Cardot's name is merely a section name it can hardly be
used in this sense.   To avoid further misunderstanding I would propose the
following binomials in which Burnettia replaces Homalothecium: **Burnettia
sericea** (L.), **B. Philippeana** (Spruce).

I would also note the error in the BRYOLOGIST 6:65, where *subcapillatum*
was written for *subcapillata* (Hedw.).   A. J. GROUT.

DIE EUROPAISCHEN LAUBMOOSE, BY G. ROTH.
Bd. 1. Lief. 1-3. Leipzig. Wilhem Engelmann. 1903.
V. F. BROTHERUS.

Since the publication of the Bryologia Europaea, that monumental work
upon European Moss Flora, several illustrated works upon the same subject
have appeared.   The Bryologia Europaea is, however, on account of its high
price less attainable and is at the present time far from complete, and the
others refer only to more or less extensive portions of Europe.   The work we
have now the pleasure to announce fills up, therefore, a considerable blank in
bryological literature by giving not only descriptions but very instructive
illustrations of nearly all the known species.   Of the few species which the
author has up to the present time not succeeded in obtaining, he hopes to
be able to give in a supplement with the necessary illustrations.

Mr. Roth's work, which will comprise two volumes, with one hundred
and ten plates, will come out in ten or twelve parts quickly following one
another.   In the general part (pp. 1-92) are treated in the most exhaustive
manner, also taking into consideration the newest literature, the anatomical
formation of mosses, their manner of increase, extension in a vertical and
horizontal direction, their relation to the substrata and importance in the
economy of both nature and mankind, as also a review of the most important moss systems.   Special interest will be taken in the author's detailed setting forth of the importance of mosses in nature, a subject which so far as we
are aware, has not been before treated in bryological literature.

In that particular part which is introduced by a copious index of the literature, the author, in all concerning the system, agrees almost entirely with Schimper. The descriptions of the systematic groups of greater or smaller extent as well as of the species are carefully and critically drawn up and are less minute than in Limpricht's well-known work, With the species, account is also taken of their synonyms and briefly of the geographical extension. Unfortunately there is wanting a "clavis" which in the genera specially rich in species would have been very useful.

The illustrations which are all from the author's own hand, and are reproduced by photo-lithography, refer chiefly to anatomical details, leaves and sporogones; but with the smaller species are also found habitat pictures. Of course it is just these drawings, executed with so much pains, that give to the work its greatest value. As regards the printing, the book is very handsomely gotten up, and the price ($1.00 each part) must be considered extremely moderate.

We congratulate the author upon this beautiful work, and are convinced that it will win for itself many friends.

Helsingfors, Finland, Sweden.

SOME ERRONEOUS REFERENCES.

J. FRANKLIN COLLINS.

Some erroneous references which appear in Limpricht's Die Laubmoose have been quoted by Prof. Holzinger in his article on page 8 of the January BRYOLOGIST. It may be well to call attention to these and also to others in connection with *Hymenostomum* which appear elsewhere. As the errors are identical—translation excepted—in both Prof. Holzinger's article and in Limpricht's work, I will refer directly to the former as the latter will probably be inaccessible to a majority of the readers of this article.

In regard to *Hymenostomum* it is stated that " The authors of the Bryologia Germanica (1823) emphasize its affinity with *Weisia viridula*." In the work mentioned, Theil I, page 191 (1823), the relationship with " *Weisia controversa*" is emphasized, while *W. viridula* is not mentioned anywhere on the nineteen pages (188-206) devoted to the genus *Hymenostomum*. Practically, this may be of little importance as *W. controversa* is now usually regarded as a synonym of *W. viridula*.

Near the middle of page 8 in THE BRYOLOGIST it is stated that in the Bryologia Europaea " *H. rutilans* (Hedw.) is made a synonym of *Weisia mucronulata* Bruch, and *H. subglobosum* Bryol. Germ. a synonym of *Weisia viridula*; further, *H. rostellatum* is treated as an *Astomum*, and by error *H. crispatum* Bryol. Germ. is also removed." Turning to the Bryologia Europaea, Fasc. 33-36 (1846), one notes firstly, that *H. rutilans* is made a synonym of *Weisia mucronata* Bryol. Eur. (not of *W. mucronulata* Br.); secondly, that *H. subglobosum* does *not* appear as a synonym of *Weisia viridula*; and lastly, that it is *Phascum* (not *Hymenostomum*) *rostellatum* which is treated, Fasc. 43 (1850), as an *Astomum*, although one of these may be a synonym of the other. The statement that "*H. crispatum* Bryol. Germ. is also removed" may be somewhat misleading, or perhaps vague. It is true that it is removed from the genus *Weisia* for it is described and figured as *Hymenostomum crispatum* Nees et Hornsch. Not all of the references in Limpricht have been verified, as several of the works mentioned were not available.

While investigating the references above I was considerably astonished to discover several errors of more or less importance on a single page of the Bryologia Europaea, and all in connection with the Bryologia Germanica. So numerous were these mistakes, though sometimes unimportant, that it almost seemed as if I possessed an edition of the work different from the one cited by Bruch and Schimper, notwithstanding the fact that a portion of the references were correct and that my copy bore the date "1823" on the title page of Theil I and also "1821" at the end of the preface.

For the sake of brevity I will quote the lines in the Bryologia Europaea —page 5 of the *Weisia* monograph, Fasc. 33-36 (1846)—which contain the errors. Each pair of brackets (here inserted by the writer and not appearing in the original) inclose the corrected reference for the immediately preceding portion.

"*Weisia humilis, W. fallax* et *W. Bruchiana* NEES et HORNSCH. *Bryol. germ.* P. II, Sect. 2, p. 36 et 38 [p. 36, 38 et 50], Tab. XXVI et XXVIII."

"*Weisia controversa var. γ stenocarpa* NEES et HORNSCH. *Bryol. germ.* p. 45 [Th. II, Abt. 2, p. 45], Tab. XVII [Tab. XXVII]."

"*Weisia amblyodon* BRID. *Bryol. univ.* I, p. 805.—NEES et HORNSCH, *Bryol. germ.* II, p. 33, Tab. XXV [Th. II, Abt. 2, p. 52, Tab. XXVIII]."

"*Weisia gymnostomoides* et *microstoma* [These two specific names should be transposed, as the first reference following belongs to the second species and the second reference to the first species] NEES et HORNSCH. *Bryol. germ.* p. 34 [Th. II, Abt. 2, p. 34], Tab. XXV et p. 52, T. XXXVII [p. 33, Tab. XXV]."

"*Weisia Rudolphiana* NEES et HORNSCH. *Bryol. germ.* II, p. 33 [Th. II, Abt. 2, p. 31], Tab. XXV."

Hymenostomum subglobosum is described on page 203 in Theil I of the Bryologia Germanica. By typographical error this page is numbered *103*. It is worthy of note that this error of paging has been copied in Bridel's Bryologia Universa 2:80 (1827), in Mueller's Synopsis Muscorum 1:651 (1849), and in Páris' Index Bryologicus, page 1,368 (1898), although it was corrected in Limpricht's Die Laubmoose 1:255 (1886).

The writer can find no authority for the specific name "*mucronulata*", mentioned above. It may be an error. The combination "*Weisia mucronulata*, Schimp." appears in an article by T. P. James in Vol. XIV (1879), p. 136, of the Proceedings of the American Academy of Arts and Sciences. Both editions of Schimper's Synopsis, however, spell the specific name "*mucronata.*" Providence, R. I.

A NEW BRACHYTHECIUM.
Brachythecium rivulare B. & S. var. tenue n. var.
A. J. Grout.

European and Northeastern American forms of *B. rivulare*, almost without exception, have the stem leaves obtuse to obtusely acute, never slenderly acuminate, but Northwestern American forms seems to. have a tendency to vary in the direction of acute to slenderly acuminate stem leaves. *B. Nelsoni* Grout is the extreme limit of this tendency that has thus far come under my observation. *B. rivulare laxum* Grout is another illustration of this tendency, and Prof. John M. Holzinger has discovered a third very interesting form in Lamoille Cave, Minnesota, to which I desire to give the name of *B. rivulare tenue.*

Prof. Holzinger's specimens were collected August 23, 1894, and have been sent out as " No. 7, Hypnum." Specimens have been in my hands for some years and the plant has been examined by M. Cardot, but neither of us has felt sure of its proper place. Recently, however, I have compared it with some specimens of undoubted *B. rivulare* from Tuckerman's Ravine, Mt. Washington, collected by myself in 1898, and find that the two can scarcely be distinguished except by the acuminate stem leaves of the Minnesota plant which is briefly characterized as follows: plants prostrate or ascending, irregularly branched, slender, very light glossy yellow, lower leaves distant and often spreading, the upper closely imbricated, giving the upper portions of the plant the appearance of forms of *B. oxycladon.* Microscopical struc-ture like that of slender *B. rivulare,* except that the stem leaves are acuminate with a rather short slender point. No antheridia, archegonia or sporophytes found.

Type in herbarium A. J. Grout. Co-types will be issued as No. 200 of my North American Musci Pleurocarpi.

MUSCI BORAELI-AMERICANI.
Fascicle I. by Prof. J. M. Holzinger.
A. J. Grout.

Every student of North American mosses will welcome Prof. Holzinger's Musci Acrocarpi Boreali-Americani as a much needed addition to the knowledge of our acrocarpous mosses. Mrs. Britton formerly planned something of the sort but pressure of other work has caused her to give up the plan.

The first fascicle of Prof. Holzinger's mosses are nearly all his own collecting, and came from an interesting section of the country. There are twenty-five numbers in the fascicle and all are abundant in material and neatly put up. As the labels do not give the name of the person determining the specimens, except in one or two cases, we presume that Prof. Holzinger is responsible for most of the naming, and the determination seem to have been made with a care that merits one's confidence. As Prof. Holzinger puts up but twenty-five sets those who intend to subscribe should do so at once or it will be too late.

THE FRUITING SEASON OF THE HAIR-CAP MOSS.

PHEBE M. TOWLE AND ANNA E. GILBERT.

Paper read before the Vermont Botanical Club, Jan. 20, 1904.

In March, 1903, a group of students in the laboratory of the University of Vermont were giving attention to the hair-cap moss. The material had been brought from different stations, and in it were found plants with the rosettes having well developed antheridia; plants with tufts of green leaves at the top, showing apparently only last year's growth; plants having sporophytes of good height and usually retaining the hairy cap; and others with the sporophytes rising only from one-half to three-fourths of an inch above the tuft of leaves. As the observations went on these questions arose: first, where are the archegonia with the egg cells which should be ready for the sperm? second, how old are these sporophytes? third, when do these sporophytes mature their spores? Following the suggestions of the questions these observations have been noted.

On March 24th the antheridial plants showed the antheridia with the contents retained, the sperm mother cells showing through the walls. April 16th the antheridia were discharging their contents. The sperm mother cells were massed together and appeared nearly square as seen in section. The motile sperm cells were in very rapid motion. On the same day an anchegonium was found. It was about as tall as the smallest or innermost enclosing leaves. On April 18th two archegonia were found in the same plant. One was about the size of the one found two days earlier and the other was two and one-half times as tall as the first. The shorter archegonium had a rounded top apparently unopened, while the taller one was somewhat funnel shaped above probably indicating the mature condition of the archegonium when it is open ready to receive the sperm cells. Many plants were examined in which no archegonia were found.

In the last week of July the hairy cap showed in a dissected specimen. It was during the third week in August that the first little hair-cap peeped out of its tuft of green leaves. One week latter the same little plant was showing plainly. On October 12th the sporophytes were showing a half-inch above the leaves. Some of these plants were brought in and kept in a cool place and two months later had added, in some plants, another half-inch to their height.

The second question,—how old are the sporophytes which are present in the spring, has been answered in part by the reports just given of the development of the plant through the season. But this will be made still more clear by observations of next spring upon plants in marked stations, of which the autumn conditions were made matters of careful record.

The third question,—when do these sporophytes ripen their spores, takes us on to August. The greater number of sporophytes of *P. juniperinum* had by August 21st shed their hairy caps, but some still retained them. The lids were in place but came off easily when disturbed and the spores could be pushed or shaken out. The sporophytes looked fresh. The capsules and spores were green in color. The plants from which they grew showed no indications of this year's growth other than the sporophytes themselves, while the other plants nearby showed the new year's growth distinctly. *Polytrichum commune* found by the roadside a little way west of Hazen's Notch on July 21st was apparently as far advanced as *P. juniperinum* found in the Missisquoi Valley on August 21st.

This year's observation would indicate that for these two species of hair-cap moss the escape of the sperm cells and the maturing of the archegonium for their reception occurs in April, and that the maturing of the spores within the sporophyte takes place one year from the following August. The early stages of the development of the sporophyte progress rather slowly. Later, in July and August, growth seems rapid. Then again in the fall growth is slower. Let those who wish to get motile sperms search in April for male rosettes in which the white tips of the antheridia may be seen just peeping out from between the scales of the rosette, if one looks carefully with a good glass. Let such plants dry slightly, then by wetting them for mounting they will show the discharge of the sperm mother cells.

When the interest in the subject began to deepen a search was made for literature relating to it. Nothing was found until the last volume of Hedwigia came out containing an article by A. Grimme: Ueber die Blüthezeit Deutscher Laubmoose und die Entwickelungsdauer ihrer Sporogone. The length of time of development of *P. commune* and *P. juniperinum* is there given, for Europe, as from thirteen to fifteen months. This corresponds closely with the observations made in Vermont. It is stated in the article referred to that no moss develops its sporophyte in less than three months, and that some take nearly two years. This shows that there is an opportunity for much interesting work in verifying this study and in finding out the life history of other of our common mosses in this country.

Botanical Laboratory, University of Vermont, Feb. 1, 1904.

NOTICE.

M. Bescherelle, whose death recently was announced in this journal, was interrupted in the preparation of an important bryological work, a "Sylloge" of all the species of mosses described by him. M. Cardot, to whom its completion was entrusted, states that it will contain 450 to 500 pages, and that it will need to be published by *subscription*. It will be possible to print the work at $5.00 a copy, provided that at least *fifty* of the minimum of 140 subscribers necessary to begin the printing can be found in the United States. I desire to announce that I will head this list, and will also receive names of other subscribers, at Winona, Minn.

JOHN M. HOLZINGER.

NOTE.

Bescherelle's proposed "Sylloge," will cost about $3.00, not $5.00 as stated. J. M. H.

SULLIVANT MOSS CHAPTER NOTES.

"A Field Day or Moss Walk" has been proposed for all Sullivant Moss Chapter members and their friends living within ten miles of Boston, Mass. To take place on Saturday afternoon, April 23rd, or, if stormy on Saturday, April 30th. All interested in such an event write to the originator for details. Address, Mr. Walter Gerritson 66 Robbins Street, Waltham, Mass.

CHAPTER NOTES CONTINUED.

OFFERINGS.

[To Chapter Members only—for postage.]

Mrs. Mary L. Stevens, 39 Columbia St., Brookline, Mass. *Brachythecium rivulare* B. & S. Collected, Gilford, N. H.

Mr. Walter Gerritson, 66 Robbins St., Waltham, Mass. *Bryum argenteum* L. c.fr.; *Climacium Kindbergii* (R. & C.) Grout: *Dichelyma capillaceum* B. & S., c.fr. Collected, Weston and Waltham, Mass.

Mrs. Sarah B. Hadley, South Canterbury, Conn. *Bryum caespiticium* L. c.fr.; *Pogonatum tenue* (Menz.) E. G. Britton, c.fr. Collected, S. Canterbury, Conn.

Prof. W. W. Stockberger, Bureau Plant Industry, Washington, D, C. *Polytrichum juniperinum* Willd. Collected, Ohio; *Fissidens decipiens* De Not. Collected, Washington, D. C.

Miss Mary F. Miller, 1109 M. street, N. W., Washington, D. C. *Cladonia rangiferina* (L.) Hoffm.; *Pogonatum urnigerum* (L.) Beauv., c.fr.: *Brynhia Novae-Angliae* (Sull. & Lesq.) Grout, c.fr. Collected, Shandaken, N. Y.

Mrs. Carolyn W. Harris, 125 St. Mark's Avenue, Brooklyn, N. Y. *Cladonia amaurocraea* (Fl.) Schaer.

Mr. G. K. Merrill, 564 Main street, Rockland, Me. *Cetraria Islandica* (L.) Ach.; *C. Oakesiana* Tuckm. Collected, Camden, Me.

WANTED

To exchange. Alpine and subalpine New England Lichens for specimens from the South, Southwest and Pacific coast. Shall be pleased to hear from anyone interested in Lichens.

Mr. G. K. Merrill, 564 Main street, Rockland, Maine.

VOLUME VII. **NUMBER 3**

※ MAY, 1904 ※

THE BRYOLOGIST

AN ILLUSTRATED BIMONTHLY DEVOTED TO

NORTH AMERICAN MOSSES

HEPATICS AND LICHENS

EDITORS:

ABEL JOEL GROUT and ANNIE MORRILL SMITH

CONTENTS

Entered at the Post Office at Brooklyn, N. Y., April 2, 1900, as second class of mail matter, under Act of March 3, 1879.

Published by the Editors, 78 Orange St., Brooklyn, N. Y., U. S. A.

PRESS OF MC BRIDE & STERN, 97-99 CLIFF STREET, NEW YORK

THE BRYOLOGIST

BIMONTHLY JOURNAL DEVOTED TO THE STUDY OF NORTH AMERICAN MOSSES
HEPATICS AND LICHENS.

ALSO OFFICIAL ORGAN
OF THE SULLIVANT MOSS CHAPTER OF THE AGASSIZ ASSOCIATION.

Subscription Price, $1.00 a year. 20c. a copy. Four issues 1898, 35c. Four issues 1899, 35c. Together, eight issues, 50c. Four issues 1900, 50c. Four issues 1901, 50c. Four Vols. $1.50 Six issues 1902, $1.00. Six issues 1903, $1.00.

Short articles and notes on mosses solicited from all students of the mosses. Address manuscript to A. J. Grout, Boys' High School, Brooklyn, N. Y. Address all inquiries and subscriptions to Mrs. Annie Morrill Smith, 78 Orange Street, Brooklyn, N. Y. For advertising space address Mrs. Smith. Check, except N. Y. City, MUST contain 10 cents extra for Clearing House charges.

Copyrighted 1904, by Annie Morrill Smith.

THE SULLIVANT MOSS CHAPTER.

President, Prof. J. M. Holzinger, Winona, Minn. Vice-President, Mrs. C. W. Harris, 125 St. Marks Avenue, Brooklyn, N. Y. Secretary, Miss Mary F. Miller, 1109 M Street, Washington, D. C. Treasurer, Mrs. Smith, 78 Orange Street, Brooklyn, N. Y.
Dues $1.10 a year, this includes a subscription to THE BRYOLOGIST.
All interested in the study of Mosses, Hepatics, and Lichens by correspondence are invited to join. Send dues direct to the Treasurer. For further information address the Secretary.

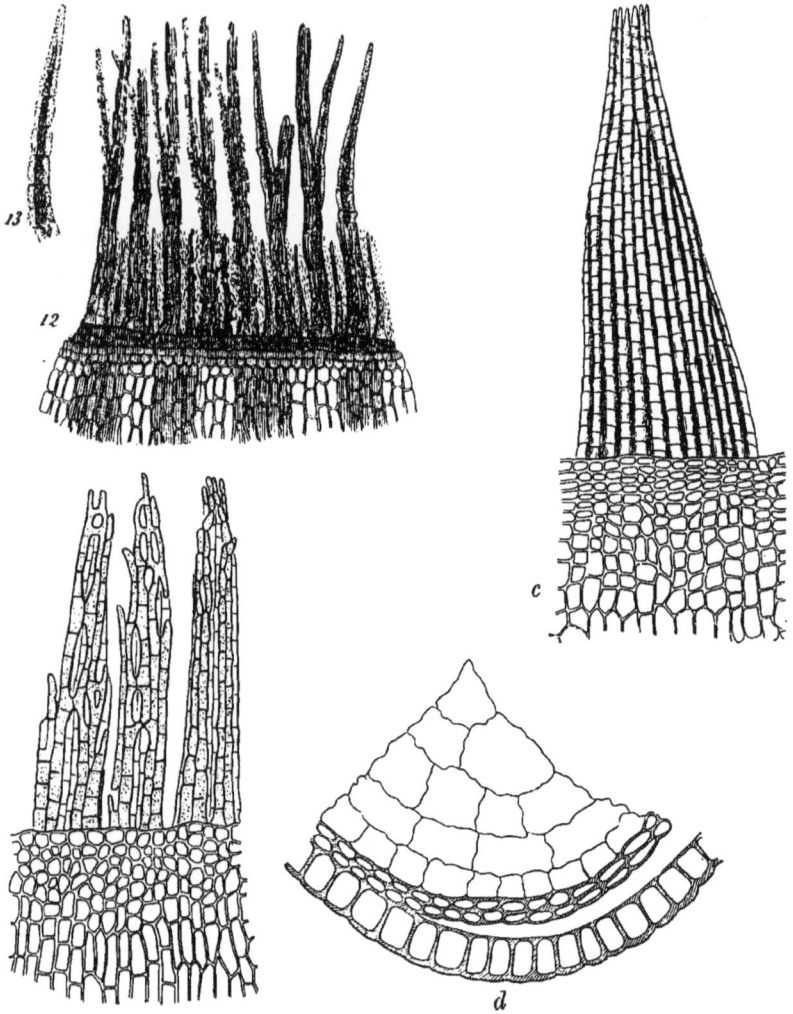

PLATE IV. Figs. 12 and 13. Peristome of *Encalypta procera*. Fig. *c*. Tooth of *Georgia pellucida*. *d*. Cross section of same. Lower left hand figure represents three teeth of *Encalypta longicolla*.

THE BRYOLOGIST.

Vol. VII. MAY, 1904. No. 3.

THE PERISTOME, VI.

A. J. GROUT.

In *Funaria* the segments are formed by the thickening of the ventral or inner walls of the sixteen cells, instead of adjoining portions of two cells as in *Mnium*; hence in *Funaria* the segments are directly opposite the teeth instead of alternating with them (See BRYOLOGIST, V., p. 6). Moreover, the exterior surface of the segments consists of a single row of plates instead of a double row, as in *Mnium*; this last follows as a necessity from the position of the segments, but on the inner side of the peristomial layer of cells, instead of several irregular rows of cells there are just two rows, corresponding in position exactly to the two rows on the outside of the peristomial layer in *Mnium*, so that the *inner* face of the segments consists of a double row of plates like the *outer* surface of the teeth.

M. Philibert suggests that the inner peristome of *Bartramia* may be intermediate between these two types, for while the segments are carinate as in *Mnium*, they are split along the keel; if now the adjoining halves of each pair of segments were to be united, we should have a condition strongly resembling that in *Funaria*. As there is a basal membrane which does not split at all this hypothesis does not seem at all forced. In *Dicranum* and most of the *Aplolepideae* the outer side of the teeth consists of a single row of plates, like the segments in *Funaria*. The median line is the line of junction of the two rows of plates which form the inner side of the tooth. For these reasons and some others not so easily explained, Philibert has concluded that the peristome of the *Aplolepideae* is homologous with the inner peristome of the *Diplolepideae*. For this reason it seems objectionable to speak of the endostome.

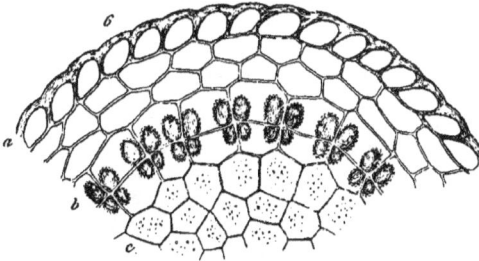

Fig. 1.

Philibert says: "Mosses which have only a single row of plates on the dorsal surface of their teeth never have a double peristome."

In *Tortula subulata* (L.) Hedw. (Fig. 1,) the peristomial layer is bounded on its inner surface by another layer of sixteen cells, exactly matching the peristomial layer in position. The teeth are formed by thickenings in the four contiguous corners of this double row of cells. There are thus sixteen separate centers of deposit made up of four parts each. The thirty-two teeth of *Barbula* and *Tortula* are made by the splitting in half radially of each of these sixteen bundles. The thickenings are thus all on the inner cell wall of the peristomial layer as is required by Philibert's hypothesis. He states that in the basal membrane of this *Tortula* the layer inside the peristomial layer consists of just twice as many rows of cells as appear in the region from which our figure is taken, thus making the homology with *Dicranum* and *Funaria* more complete.

Returning now to the relations between the arthrodont and nematodont types, M. Philibert considers *Encalypta* a primitive form connecting the two types and giving rise to all the varied kinds of arthrodont peristomes. In this connection I am very forcibly struck with the leaf resemblances between *Georgia, Webera, Encalypta* and the *Tortulaceae*, which last are evidently next of kin to *Encalypta* on the arthrodont side. Moss students will remember that so eminent an authority as Braithwaite places *Webera sessilis* next the *Tortulaceae* because of its leaf structure.

The leaves of *Buxbaumia* are so reduced as to be of little value in showing relationships, but the peristome shows undoubted affinity to that of some species of *Encalypta*. My studies on the peristome have convinced me that the following arrangement of families would more truly represent the order of evolution than the one I have previously followed: *Georgiaceae, Polytrichaceae, Buxbaumiaceae, Encalyptaceae, Tortulaceae, Ephemeraceae, Grimmiaceae, Dicranaceae, Fissidentaceae.*

In *Encalypta* we find a curious combination of peristome characters. The peculiar "extinguisher-like" calyptra and the leaf characters make it certain that no mistake is made in putting all the species into one genus, yet within the limits of this genus we have almost all degrees of completeness of the peristome from none at all in *E. commutata* to simple in *E. ciliata*, and highly developed and double in *E. procera*. The relationship to the Nematodont peristome is shown in *E. longicolla* and *E. brevicolla*, especially the former. Comparing the figures of the peristome of *Georgia pellucida* and those of the peristome of *E. longicolla* (Plate IV.) a most striking superficial resemblance appears, and this seems to increase rather than decrease upon closer investigation. The structure of the peristome of *Georgia* has previously been carefully explained, the only new points brought out are the thickening of the cell walls of the outer rows of cells. The peristome of *Encalypta longicolla* consists of sixteen lanceolate teeth, each composed of a bundle of jointed, red, papillose filaments. On the outer face of each tooth one can count four or five of these filaments, cohering at their joints but more or less free in the intervals between the joints, so that the tooth appears somewhat perforate. Sometimes these filaments are divided irregularly into two groups, separate at base but united at the top. These filaments contain cellular cavities as in *Georgia*, two in the middle and thickest portion of the tooth but usually only one on the edges. These cell cavities are enclosed by walls composed of two layers of plates, as are the teeth of the Arthrodonteae. "To pass to the normal type of the Arthrodonteae nothing is necessary but a reduction in the number of elements of the peristome and a more complete and regular separation of its parts. The number of the teeth, so variable in the *Buxbaumiaceae* and also in the *Polytrichaceae* is already reduced to sixteen in our *Encalypta*, but each of these yet contains several layers of cells in thickness and several rows in width." The gradual reduction to the normal type takes place almost before our eyes in the genus *Encalypta*. There are forms of *E. longicolla* in which there are but three filaments on the face of each tooth, and there is often only a single layer of internal cell cavities. The reduction becomes even more apparent in the closely related *E. brevicolla*.

In *E. apophysata* the peristome is composed of sixteen long and narrow teeth connivent in the form of a cone. Upon the dorsal face of each of these teeth are two rows of reddish or orange plates separated by a median line along which the tooth is sometimes perforate. In thickness there are ordinarily three layers, two in contact with each other and a third inside these and separated from them by narrow elongated cell cavities. This layer rarely extends the whole length of the teeth and may be present near the base only. It is easy to recognize on the one hand the homologies of this peristome with that of *E. longicolla* and on the other with the normal Arthrodont peristome. The two outer layers of united plates correspond to the two layers of plates in the teeth and the third corresponds to the inner peristome.

In *E. procera* (Plate IV. Fig. 12,) there is an external peristome of sixteen narrowly linear, much elongated teeth, but which in number and arrangement of plates is typical of the Arthrodont-Diplolepid external peristome, i. e., two rows of outer plates and one of inner plates. Directly inside these outer teeth are sixteen inner teeth. These inner teeth are a little shorter than the outer and are formed of an outer papillose layer of plates and an inner more strongly thickened layer. These inner teeth are united below into a basal membrane about one-fourth the entire height of the inner peristome ; this membrane is also united to the outer teeth by unabsorbed radial walls as in *E. apophysata*. Alternating with the teeth are sixteen narrow processes which Philibert states to be homologous with the keels of the inner peristome of the Diplolepideae.

Comparing this peristome of *E. procera* with the typical peristome of the Diplolepideae as illustrated and explained in previous articles there will be little difficulty in recognizing the corresponding and homologous parts. In the later development in *Mnium* and *Hypnum* the reduction in the number of elements in the teeth has been followed by a great increase in the breadth of the remaining elements.

AN INTERESTING MOSS BOOK.

A. J. GROUT.

Mr. Wm. L. Sherwood, President of the New York Naturalists' Club, has in his library a copy of "Twenty Lessons on British Mosses," by Wm. Gardener, of Dundee, published by Longman, Brown, Green & Longmans, London, in 1847. It is an exceedingly interesting little volume and apparently rare as I have never before seen a copy, and do not remember ever having heard of the book.

The book is of interest chiefly for two reasons: First, it has the modern nature study idea in all its fullness. In the preface he states, "Were the youthful mind more generally directed to natural objects, and to the wonderful operations continually going on around us, and taught to seek entertainment and instruction in them, rather than in matters of a frivolous or vicious tendency, it is impossible to say to what extent this might be conducive to the future advancement and well being of human society." The chief unfavorable criticism seems to be that he preaches too directly as when he says in his introduction, "God made the little moss as well as the glowing flower and lofty tree, and has made nothing in vain." Although such sermons are frequent they are free from cant and do not seem as incongruous as they ·would in a modern book of nature study.

The second point of interest is that the book is illustrated with actual specimens instead of engravings, and these are inserted in an exquisite manner. One can still see the peristome with the naked eye on the capsule which is pasted in to illustrate the "peristome;" for the book has a short introductory lesson on structure illustrated with specimens. It is noticeable that instead of taking up twenty of the most common mosses and describing them he selects twenty illustrating pretty nearly all of the families of mosses as then recognized, and does not always take the most common or easily accessible, as when he uses *Polytrichum (Pogonatum) urnigerum* to illustrate the hair-caps, and *Weissia nigrita (Catoscopium nigritum)* to illustrate the Weissias.

When writing of mosses closely related to those illustrated he does not usually tell how to distinguish them, but merely mentions their occurrence and sometimes where he has collected them. This is not because of inability for he makes these distinctions in an admirable manner when writing of variety and species and illustrating these terms. The book (of about fifty pages 4x7 inches) was evidently well received for Mr. Sherwood's copy is of the third edition. This notice is written to call attention to one of the early attempts at nature study methods in the field of Bryology, which is well worth the consideration of any student of the development of educational ideas and ideals.

A NOTE.

Mr. J. F. Collins, of Brown University, calls my attention to a misstatement in my article on Vermont Mosses in the January BRYOLOGIST, where I state, page 7, that "*Heterocladium squarrosulum* has not before been reported from the eastern United States except Mt. Washington." It is reported from Massachusetts in Hitchcock's report of the Geology, Botany, etc., of Massachusetts (1833), page 640, and in Tuckerman & Frost's Amherst Catalogue (1875), and in Cobb's List of Plants near Amherst (1887). The last two are undoubtedly based on Frost's specimen collected in Brattleboro, Vt., in 1853, which specimen I have examined through the courtesy of Prof. L. R. Jones, of the University of Vermont. It is not *H. squarrosulum* at all but an entirely different moss, but on the margin is pencilled, "First collected in this country by me. C. C. F." The 1833 report still remains to be verified or disproved.   A. J. GROUT.

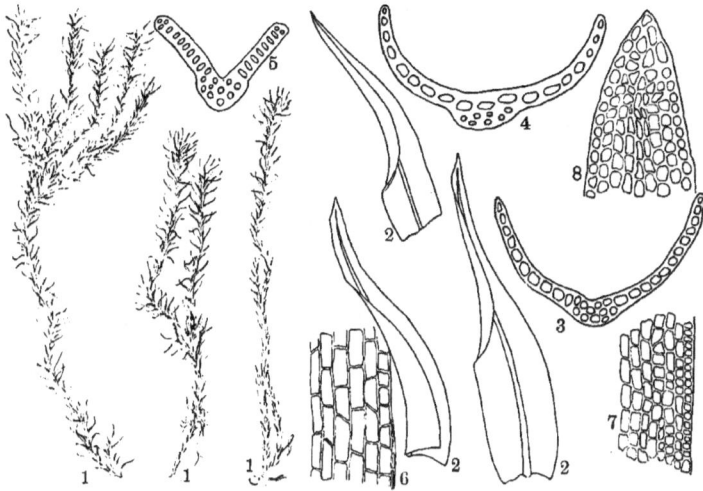

Rhacomitrium FLETTII.

PLATE V. Fig. 1. Plants showing mode of branching and disposition of leaves
when moist. Natural size about 1.5 cm. × 3. Fig, 2. Leaves × 10.
Figs. 3, 4, 5. Cross sections of a leaf from near base, middle and apex,
respectively. Figs. 6, 7, 8. Leaf areolation from near base, middle and
apex, respectively.

RHACOMITRIUM FLETTII n. sp.

JOHN M. HOLZINGER.

Stem densely caespitose, radiculose at base, simple or fasciculately
branched; color of plants yellowish brown. Leaves divergent, then ascending when moist, lanceolate, margin entire; costa reaching apex; cells at base
pellucid, thin-walled, approximately rectangular, about 1x2, above more isodiametric, thick-walled, with a row of more pellucid, smaller, roundish cells
along the margin and this bistratose toward the apex. Entirely sterile.

This plant was collected by Prof. J. B. Flett on Mt. Tacoma (Mt. Rai-
nier), at an altitude of 14,519 feet, near one of the steam jets issuing from a
fissure in the side of the rim of the crater, Aug. 10, 1897.

In appearance, color and branching it looks like a diminutive form of
Rhacomitrium ellipticum, section Dryptodon. But it has almost completely
lost the very unequal thickenings of the cell walls above the base, has narrower leaves lacking the foldings in that species, has a heavier costa, and
has a tendency to a doubling of the small marginal cells, as is shown in Fig.
5 of the accompanying plate.   Winona, Minn.

A BRYOLOGIST'S GLIMPSE INTO GEOLOGICAL HISTORY.

JOHN M. HOLZINGER.

That even 'the most modest efforts and observations may contribute important data to the illumination of profoundly scientific problems, is shown in the following letter from Dr. Emilio Levier, a physician in Florence, Italy, and an enthusiastic bryologist. Dr. Levier makes so vivid a statement of the case and has so amiably permitted its use, that I feel justified in giving it in full.

"Extraordinarily interesting to me was your discovery of the identity of the Cis-Caucasian *Leskea crassiretis* with the North American *Fabroleskea Austini* (See BRYOLOGIST, Sept. 1903). If I had known of this earlier I could have communicated to you two similar cases of widely separate localities which are very suggestive not only from the standpoint of phyto-geography but also of geological history.

"The first case concerns, to be sure, not a moss, but a parasitic fungus which I collected in 1890, first at Trapezunt, later in western Trans-Caucasia (Svanetia) on *Rhododendron flavum*, and which I sent for determination to Prof. Magnus, of Berlin. Mr. Magnus believed at first he had a new species before him, entirely different from the European *Exobasidium rhododendri* Cramer. However, he recognized very soon the complete identity of the fungus from the Caucasus with the North American *Exobasidium discoideum* Ellis, which grows in New Jersey on *Rhododendron viscosum*. In Europe there is at present no longer found any species of Rhododedron of the section *Azalea maxim. (Pentanthera* Don.). The Caucasian and North America Rhododendron are genuine, closely related Azaleas. Prof. Magnus then says this, in Sommier et Levier, Enumeratio Plant. Caucas., p. 541: 'These host plants. are relics from tertiary geological time, during which North America and Europe still constituted one common floral region. While the host plants in the two very distant areas which lay formerly within one common circle of distribution, were differentiated into two closely related species, the parasitic Exobasidium has maintained itself unchanged, at least in its external morphological characters. *Exobasidium discoideum* Ell. is a parasitic fungus which inhabited the parent forms of *Rhododéndron viscosum* (L.) Torr. and of *R. flavum* Don. since the time when North America and Europe still formed one continuous floral region.'

"In the two cases of Leskea and Exobasidium, Europe is completely overleaped and the two widely separate localties stand without connecting area: Caucasia-North America. (According to Reclus and many other geographers Cis-Caucasia to the south of the Manitch depression belongs to Asia and thus also the stations of *Leskea grandiretis*, but this in no wise alters the main point at issue.)

"Now we have a third case where a trace of an intermediate locality connecting the Caucasus with North America has been established by the discovery of à relic, no trace being left in all the rest of Europe. The case is that of a moss, *Mnium ciliare*. This moss grows and fruits abundantly in western Trans-Caucasia. I gathered it in Svanetia in flower; Brotherus in fruit. The species is given as rare in Scandinavia and from there leaps over to northern North America, is however, absent in the Alps and in the rest of Europe. Should the Scandinavian station be destroyed, the case of *Mnium ciliare* will be exactly identical with *Leskea Austini* and *Exobasidium discoideum.*" Thus Dr. Levier.

His closing words characterize fairly well the typical moss student's attitude. "Bryology has so far furnished precious little toward the evolutional history of the world floras; specialists are so absorbed in microscopic work and 'areolation' that they find no time to look beyond."

A note by me in the Plant World, Oct., 1901, showing the unusual distribution of *Claytonia chamissoi, Coscinodon Raui, C. Wrightii, Ditrichum flexicaule brevifolium, Grimmia teretinervis, Webera proligera* and *Weisia Wimmeriana:* another note in the Botanical Gazette, Aug., 1900, establishing the occurrence in the difficultly accessible mountains of N. W. Montana of several European alpine species such as *Grimmia mollis, G. subsulcata, Hypnum fluitans brachydictyon, H. ochraceum uncinatum, Webera carinata:* the report on another moss from this Montana collection in the BRYOLOGIST, April, 1901, viz. *Hypnum Bestii,* which according to M. Renauld occurs in but slightly modified form in the Pyrenees (var. *Pyrenaicum* Ren.): a note in Asa Gray Bulletin, Oct., 1900, on the occurrence of *Polytrichum Jensenii* Hagen, in the Yellowstone National Park: and a brief note on some mosses new to Alaska in the BRYOLOGIST, March, 1902, p. 30 —seem now to have an added interest. Winona, Minn.

A CALL FOR ASSISTANCE.

It is my purpose to make as complete as possible the series of Acrocarpi Boreali-Americani which I have begun to issue the first of this year, and of which three fascicles, Nos. 1-75, are distributed. I ask for the co-operation of all moss students in the various parts of the United States. I shall offer suitable remuneration to any experienced collector, either in mosses or cash, for sets of forty to fifty good specimens of such species as are outside of my reach. And I invite correspondence for this purpose.

After the first century of this series is complete, I propose to publish a pamphlet summing up the criticisms and corrections, which I herewith cordially elicit from all recipients, and it is my intention to publish with this series of notes a full list of all recipients with their addresses, believing that this will be an aid in the study of the mosses distributed, and in establishing a uniform understanding of critical and little understood species.

May I be permitted to point out, in closing this note, that I am taking very great pains to avoid all mixtures. My task will be somewhat simplified if those collecting for me will use due caution not to mix sods in the field. JOHN M. HOLZINGER.

NOTE.

I find that *Weisia mucronulata* Bruch is described on page 158 in Vol. IV, Part I of Sprengel's Linnæi Systema Vegetabilium (1827). The first eighteen words of the last paragraph of my article in the March BRYOLO-GIST consequently become irrelevant, and the reference to " *W. mucron-atula* " in Paris' Index Bryologicus (p. 1365) under *Weisia mucronata* is shown to be incorrect. At the same time I would suggest that the word "quoted " in the first paragraph of my article be changed to *translated*.
March 14, 1904.   J. FRANKLIN COLLINS.

SUPPLEMENTARY NOTE.

Owing to the thoughtless omission of a certain sentence from the first paragraph of my article in the March BRYOLOGIST (which paragraph was hastily condensed just before going to press) a wrong impression may have been conveyed to some of the readers of the article—an impression which would be unfair to Prof. Holzinger. In order to counteract any such possible impression I wish to say that Prof. Holzinger cannot justly be held for the errors cited, because they appear in his stated source of information, and, furthermore, they were, in his article, inclosed within quotation marks. His purpose, which is seen to be fully served by an exact translation, and my own in calling attention to certain erroneous citations, have no direct bearing whatever upon each other. My object in citing portions of his article was simply to make my own more intelligible by referring directly to an article which was accessible to all.

Prof. Holzinger suggests that " is also removed " which appears on page 8 (line 22) of the January article be changed to "is retained " (i. e. as a *Hymenostomum*).   J. FRANKLIN COLLINS.
March 21, 1904.

A CORRECTION.

Thanks to Mr. Collins' note on Some Erroneous References in the March BRYOLOGIST, I am able to make an important correction in my translation: the word "fortgeführt," rendered by me "removed," must be rendered "retained in the genus."

Then I also thank Mr. Collins for his correction in the above note substituting "translated " for "quoted." It was entirely foreign to my purpose to criticise Mr. Limpricht's page references; the simple translation was all I needed, and the references, erroneous or right, had no bearing on my point.
JOHN M. HOLZINGER.

Mr. J. F. Collins has called my attention to the fact that Dixon has named a *Brachythecium rivulare tenue* so that my name is untenable. For the variety described in the April BRYOLOGIST under this name I propose the name **B. rivulare LAMOILLENSE**.   A. J. GROUT.

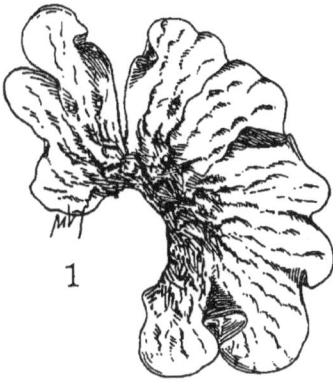

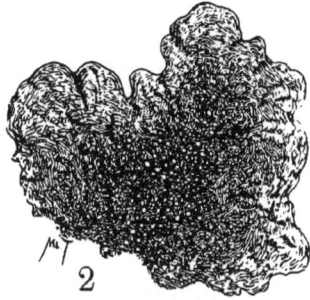

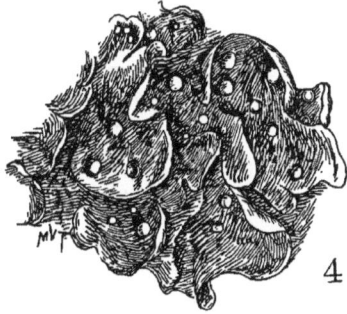

PLATE VI. Fig. 1. *Collema flaccidum* × 1. Fig. 2. *C. nigrescens* × 0.
Fig. 3. *Leptogium palmatum* × 2. Fig. 4. *L. pulchellum* × 2. Fig.
5. *L. myochroum* × 0.

LICHENS—COLLEMA AND LEPTOGIUM.

CAROLYN W. HARRIS.

Collema is a large genus, represented by many species which after further study may be divided into several genera, as the spore characters are such that this division seems possible. The thallus is foliaceous, in a few species fruticose, in some it is small and irregular, in others large and flat, but in all it is more or less folded and rugose. Collemas are very gelatinous when moist, very thin in texture. They are generally considered of low development but they have some characteristics which are quite complex. The color varies from a light lead color to a very dark, greenish blue, almost black, the under surface is lighter with no rhizoids. The margin of the thallus is divided into lobes with obtuse tips. The apothecia are usually small or medium, frequently very numerous. The color of the disk is reddish brown to dark brown, flat or slightly convex with an entire, crenate margin. Collemas are not dependent upon apothecia for propagation as soredia are usually present, and it is even possible for a part of the thallus to be broken off and form a new plant. They are found on trees, dead wood, occasionally on damp rocks. They require much moisture for their full development. Are common throughout the United States and Canada.

Leptogium. The thallus of this genus closely resembles that of Collema; it is usually larger and not so gelatinous, more wrinkled and thicker. On the under side are fine gray rhizoids in clusters, or there is a fine close nap. The apothecia are small with a reddish brown disk, with a lighter raised margin. Soredia are present in many species where no apothecia are found. Leptogium is probably only a higher development of Collema; the two genera are found in the same localities. The amateur can distinguish these genera readily by examining the under surface, in Collema it is devoid of rhizoids, in Leptogium they are always present in some form.

COLLEMA FLACCIDUM Ach. Plate VI. Fig. 1. Thallus composed of large, round, entire lobes which are folded together, giving a puckered effect; these are usually covered with granules of the same color as the thallus which is a dark olive green. The margin is undulate and frequently plicate or folded. The under side is a dull dark gray, much wrinkled.

Apothecia, which are not common, are small, scattered and sessile, the disk is chestnut, quite flat, with an entire, granulate margin.

Habitat granite rocks and trunks of old trees; found principally in mountainous regions. When moist is very flaccid as the specific name indicates, when dry it is thin and brittle.

COLLEMA NIGRESCENS (Huds.) Ach. Plate VI. Fig. 2. The thallus of this very pretty lichen is thin, nearly round, very delicate at the margin but stronger toward the center where the small apothecia are massed. The lobes are entire, round and flexuous, clinging closely to the substratum, but when damp can be removed with care. The wrinkles in the thallus are very conspicuous giving a pustulate appearance. It is a dark olive green turning darker when fully developed; the under side is a little lighter in color with deep pit-like radiating wrinkles.

The apothecia are small and crowded, with a deep red disk which has a thin entire margin.

Found on trees and decayed wood: quite common throughout the Northern States.

COLLEMA NIGRESCENS (Huds.) Ach. var. LEUCOPEPLA Tuckerm. This variety is more common than *C. nigrescens*, especially in the Southern States. The thallus, which is orbiculate, is a little thinner and more wrinkled, but the chief macroscopical difference is in the apothecia which are very small and crowded, with a convex disk which is white pruinose, giving a very attractive appearance with the often perfect outline of the thallus.

COLLEMA PULPOSUM (Bernh.) Nyl. Thallus medium in size with crowded lobes which are crenate at the margin, rather thick and frequently granulose toward the center. It is a dull dark green and is very gelatinous when damp.

The apothecia are large and flat with a deep orange colored disk, the margin is raised, is very thick and almost black.

C. pulposum is found imbedded in moss in calcareous earth, or on limestone rocks; more common in the Western than in the Eastern States.

LEPTOGIUM MUSCICOLA (Sw.) Fr. Thallus very minute and moss-like, the interwoven branches are irregular and much branched. In color it is almost black, with a varnished look.

The apothecia are rather large with a flat brown disk which is somewhat appressed, the margin is thin and entire.

This interesting little lichen grows over mosses and is difficult to separate from them. It grows in mountainous regions, both in the East and West, but is more common on the Pacific coast.

LEPTOGIUM LACERUM (Sw.) Fr. Thallus rather small with rounded lobes which are much wrinkled and crowded, and whose margins are very irregularly cleft into jagged, fringe like edges. The color of the thallus is a dark dull brown when dry, a dark lead color when moist; the under surface is paler with hair-like rhizoids with which it clings closely to the substratum.

Found on rocks, frequently with mosses; common in the Northern and Middle States.

LEPTOGIUM PALMATUM (Huds.) Mont. Plate VI. Fig. 3. Thallus medium with narrow, convolute lobes which have obtuse tips; these are more or less wrinkled and pitted, the edges are fringe like. The color is a deep reddish brown, with a trace of the lead color, found in all species of Leptogium.

The apothecia are very small and sessile, somewhat concave, the disk is a dark red with a light red entire margin. *L. palmatum* is found on earth growing with mosses in British Columbia and the Western States.

LEPTOGIUM PULCHELLUM (Ach.) Nyl. Plate VI. Fig. 4. Thallus medium with thin round lobes which are somewhat wrinkled and plicate; is a dark lead color with a greenish tinge when moist. It clings closely to the substratum except at the edges where it curls over slightly. The under side is a little ligher in color, is pitted and wrinkled.

The apothecia are small to medium in size, often very numerous, are sub-pedicellate with a dark reddish brown disk and a paler margin.

Habitat trees and rocks; very common in Canada and the United States.

LEPTOGIUM TREMELLOIDES (L. fil.) Fr. The thallus of this species is much like that of *L. pulchellum*, it is larger, with round, smooth, entire lobes which become crisped toward the center, and are covered more or less with concolorous granules, sometimes these are minute lobules. The color is somewhat lighter than that of *L. pulchellum*. The under side is the same color, and is wrinkled slightly.

The apothecia are medium, disk a dark red, which becomes convex, the margin is very thin.

Found in the Northern and Middle States on mossy rocks and on trees.

LEPTOGIUM MYOCHROUM (Ehrh., Schaer.) Tuckerm. Plate VI. Fig. 5. Thallus rather large with broad, flat lobes which are very coriaceous and entire, often sooty looking toward the center. In color a dull green when fresh, turning mouse color when dry. The lobes of the under side are slightly concave and are a light gray, covered with a fine ash colored nap.

The apothecia are rare; they are medium and flat, almost sessile, with a reddish brown disk, the border rugose and sometimes hirsute.

Found on trees and damp rocks; quite common in the Northern States.

In appearance *C. flaccidum* and *L. myochroum* are much alike, but the under side of the former is bare and in the latter it is velvety.

NOTES ON NOMENCLATURE III.

ELIZABETH G. BRITTON.

Brachelyma robustum (Cardot).

Cryphaeadelphus robustus Cardot. (Rev. BRY. **3**:6-8, 1904). M. Cardot has recently described this new species collected by R. M. Harper in Georgia and referred *Brachelyma subulatum* Sch. to the same genus.

He says of this new generic name that Müller had created it in 1851 in the second volume of the Synopsis Muscorum for *Fontinalis subulata* P. Beauv. as a subsection of the section *Dichelyma* of the genus *Neckera*. In the second edition of the Synopsis Muscorum Europaeorum, 1876, Schimper founded the genus *Brachelyma* for the same species. In his monograph of the *Fontinalaceae* in 1892, M. Cardot took up *Brachelyma*, but he states that this is an error, as according to the Paris Code, section 58, any subdivision of a genus takes rank over a later published generic name.

This name, besides being much less desirable than *Brachelyma*, is entirely misleading in its suggestion of relationship and M. Cardot renders himself particularly liable to ridicule in view of the numerous sarcastic paragraphs published by him on nomenclature in his Revision of the types of Hedwig!

PAPILLARIA NIGRESCENS (Sw.) Jaeg. & Sauerb.

In the BRYOLOGIST for March (1904) M. Cardot has failed to note that in the January number (p. 14) Louisiana was included in the range for this species. Through the kindness of Prof. Wittrock, of the Botanical Museum at Stockholm, a fragment of the type from the Herbarium of Olaf Swartz, collected in Jamaica, was sent for comparison with Florida specimens. They represent the slender terminal branches of a specimen which Swartz says was a foot long. The branches are terete, as described, the leaves appressed, about 1 mm. long, acuminate with subulate points .1-.2 mm. long, and there are 2–5 papillae on each cell.

Papillaria nigrescens var. Donnellii (Aust.)

This variety differs in its bright green or yellow color, smaller size, more slender branches, many of the terminal ones being denuded at apex or ending in a leafy tip, leaves lanceolate-acuminate, cells papillose. There is but one fruiting specimen in the Austin Herbarium and that was collected in Louisiana by Dr. Charles Mohr. It has a seta 5 mm. long with conspicuous exserted paraphyses. The capsule is old and the peristome gone.

Some Florida specimens have shorter, less acuminate leaves with larger, less papillose cells. They are generally black and stouter, with short terete branches, and closely appressed leaves, and may be referable to either the var. *brevifolia* Hpe., type locality Brazil, or to the var. *illecebra* (Brid.) C.M., type locality Haiti, but no specimens of either of these exist in our herbarium Mr. Nash collected some specimens in Haiti (851) which are stouter, larger and darker colored than any from Florida. Specimens collected by Glaziou (7393) in Brazil show a marked difference between the branch-leaves and those of the stem, both in size and shape, and may be referable to *P. appressum* C. M.

Pilotrichella cymbifolia (Sull.) Jaeg.

This name is wrongly cited by M. Cardot on p. 30 of the BRYOLOGIST, 1904, it should be given as above. It may be of interest to note that this species has clusters of septate gemmae in the axils of the leaves and evidently reproduces in this way, as the fruit is unknown.

Pilotrichella Ludoviciana (C. M.) Jaeg.

I cannot agree with M. Cardot in reducing this species to *P. cymbifolia*, (Musci Am. Sept, 44. 1893.) as it appears to be more regularly pinnate, the leaves not decurrent with fewer quadrate alar cells than in *P. cymbifolia*, the alar cells forming a brown auricle, leaves longer and more acuminate and less distinctly serrate and papillose. No. 78 R. C. Musci Am. Sept. Exsicc. appears to be *P. Ludoviciana*. Austin described the fruit (Bot. Gaz. 4:161, 1879). Nor can I agree with Kindberg (Br. Eu. & N. A. 1:15, 1897) in referring both these species to *Pterobryum*.

Pilotrichella Floridana (Aust.).

Neckera (Pilotrichum?) Floridana Aust. Bot. Gaz. 4:152, 1879. Specimens of this species cannot be found in Austin's Herbarium, but from the description it would seem to differ only from the two preceding in the simple or sparsely branched stems and almost entire minutely papillose leaves. It may be the young stages of *P. Ludoviciana*.

Ectropothecium Caloosiense (Aust.).

Hypnum (Rhynchostegium?) Caloosiense Aust. Bot. Gaz. **4**:161, 1879.
This species was doubtfully referred to *Rhynchostegium* by Austin, and was placed with the species "insufficiently known and not certainly referable to this subgenus" by Lesquereux & James. Recent studies of these genera have convinced me that it undoubtedly belongs to *Ectropothecium* in the section with the leaves having long linear cells near *E. globitheca* (C. M.). Dr. Small has recently collected three species of this genus in Southern Florida.

HOMALOTHECIUM AND BURNETTIA.

In the discussion between Messrs. Cardot and Grout several synonyms have been overlooked, notably *Pleuropus* Griff. 1849, not Gray, 1821! The following are also synonyms:

Pterigynandrum subcapillatum Hedw. Spec. Musc. 83, t. 16, 1801.
Pterogonium subcapillatum Schwaegr. Suppl, **1**:1, 107, 1811.
Pterogonium decumbens Schwaegr. Suppl. **2**:1, 32, t. 110, 1823.
Lasia subcapillata Brid. Bryol. Univ. **2**:202, 1827.
Pterigynandrum brachycladon Brid. Bryol. Univ. **2**:185, 1827.
Pterogonium ascendens Schwaegr. Suppl. t. 243, 1828.
Hypnum subcapillatum C. M. Syn. Musc. **2**:352, 1851.
Homalothecium subcapillatum Sull. Mosses N. A, 63, t. 5, 1856.
Myrinia subcapillata Kindb. Can. Rec. Sci. **21**: 1894.
Platygyrium brachycladon Kindb. Can. Rec. Sci. **21**: 1894.
Helicodontium subcapillatum Kindb. Br. Eu. & N. Am. **1**:27, 1897.
Homalothecium (Homalotheciella) subcapillatum Card. Bull. Herb. Boiss. **7**:374, 1899.
Burnettia (Homalothecium) subcapillatum Grout Bryol. **6**:65, 1903.

In my opinion *H. subcapillatum* is generically distinct from *H. sericeum* and *H. Phillipeanum*, and M. Cardot is right in calling it *Homalotheciella* according to the Paris Code. Dr. Grout has cited (Homalothecium) in parenthesis as if it were a section of *Burnettia*, and according to the new Philadelphia Code his combination is "*incidental*" and *incorrect!* Even in the last BRYOLOGIST, M. Cardot's name has priority of place, and Dr. Grout has used an American name for two European species.

New York Botanical Garden.

THE SPECIFIC (?) VALUE OF THE POSITION OF THE REPRODUCTIVE ORGANS IN BRYUM.

A. J. GROUT.

I have for a long time been of the opinion that the genus Bryum contains far too many species based chiefly on the position of the antheridia and archegonia. From the very nature of the case, if we accept the teachings of the theory of evolution, these characters must at some period in the development of the genus have been variable, e. g , at the time when dioicous species developed from monoicous ancestry there must have been a period when the species was imperfectly dioicous before it could become completely and fixedly dioicous. There are several well known cases in the flowering plants where the same species may be either monoicous or dioi-cous. And in at least one species, *Arisaema triphyllum*, the same individual varies from season to season.

In the Revue Bryologique for January, 1904, M. Corbière, in an article entitled, Sur quelques Muscinées de Maine-et-Loire, gives striking confirmation of my previously formed opinions. He states that he himself, as well as Schimper, De Notaris, Husnot, Boulay, Braithwaite and Limpricht, has recognized that *Bryum pallescens* might be either monoicous or dioicous. But studies on the plants from the region |mentioned in his title brought to light some exceedingly interesting facts in addition. The first plant studied had three inflorescences, the terminal was exclusively female, at the base were two equal branches, one of which was exclusively male and the other mixed, synoicous! Another plant had at its summit a mixed inflorescence and of its three branches two were male and the other mixed.

Here is an interesting field for our Sullivant Chapter members. Examine hundreds of specimens of mosses of some common species for variations in the arrangement of archegonia and antheridia, keep a record and report to us the results. I am confident that such variations will be found in sufficient number to "reduce" several species.

SOMETHING NEW ABOUT BUXBAUMIA.

Early in March, Miss E. B. Brainerd, Secretary of the New York Naturalists' Club, sent me two fine healthy plants of *Buxbaumia aphylla* with a large lump of soil in which they were growing. As they were the first fully developed fresh capsules I had ever seen I put them in the window and kept the soil moist, hoping to see the opercula fall and test the theories of spore dispersal. But, alas, one morning when I went to visit my pets they were gone. The earth was there but the Buxbaumia was not. The conditions were such that only one answer could be given. MICE having tired of our table delicacies had dined off Buxbaumia.   A. J. GROUT.

NOTES ON RARE OR LITTLE KNOWN MOSSES.

ALSIA ABIETINA Sulliv.

Recently I received a letter from Prof. Röll, of Darmstadt, enclosing a bit of *Alsia abietina* in fruit, which he collected some time ago at Tacoma, Wash. He found it upon an oak tree. In an article in Hedwigia he spoke of having collected it here but he omitted mentioning that it was in fruit. From an article I published in the BRYOLOGIST (6:3, 1903) it might be inferred that the moss did not fruit here. It is a pleasure to correct any wrong impression which my article may have created, and I wish to thank Prof. Röll for his interest in the matter.   DR. JOHN W. BAILEY,
                                                             Seattle, Wash.

BUXBAUMIA APHYLLA L.

While collecting along Crum Creek, Delaware Co., near Philadelphia, Pa., on April 27, 1902, I came unexpectedly upon a small colony of *Buxbaumia aphylla*. I overlooked them at first, taking them for the bud scales of the beech tree (*Fagus Americana* Sweet) which their capsules resemble somewhat at a glance. They were growing on the northern exposure of a sparsely wooded hillside, the capsules pointing up-hill towards the south. The ground was covered with a blackish film, in fact the whole situation was similar to that described by Durand in the BRYOLOGIST, **4:2**, 1901. I visited the locality again last year and saw them still growing and with an increase in number. Dr. James, in his list of the mosses of Delaware Co., gives *B. aphylla* L., but does not mention the exact locality. I would be pleased to hear from any one who could give more information as to stations in this county. EZRA T. CRESSON, JR.,

Swarthmore, Pa.

HYLOCOMIUM TRIQUETRUM BERINGIANUM Card. & Ther.

Mrs. E. G. Britton has called my attention to an extension in range for the above moss. Specimens sent for determination collected by Prof. Thomas A. Bonser, were from "Swan Range, Kootenai Mts., Flathead Co., Montana. Aug. 3, 1902. In a marshy spot of an elevated mountain ravine." "Skykomish Cascade Mts., Washington. In a low marshy spot in valley of Skykomish River." The ecological conditions were apparently favorable to its development at these points, growing mixed with Sphagnum and in a cool, wet location, at an elevation which would compensate for the habitual northern range and account for its southern migration.

BRYUM PROLIGERUM (Lindb.) Kindb.

See the BRYOLOGIST, **4**: Jan., July, Oct., 1901; also **5**: Sept., 1902. On June 10, 1903, the bed of *B. proligerum*, at Chilson Lake, Essex Co., N. Y., had extended from its original size of last year, of a foot square, to about three feet or fully three times its surface. It began to fruit June 26th, and continued to develop both gemmae and capsules till the end of the observed season, Sept. 20th. By this time it had extended along the bank for a distance of six or eight feet. A. M. S.

OFFERINGS.

[To Chapter Members only. For postage.]

Mrs. Sarah B. Hadley, South Canterbury, Conn. *Anomodon rostratus* (Hedw.) Schimp. Collected South Canterbury.

Mr. Severin Rapp, Sanford, Orange Co., Florida. *Plagiothecium micans.* (Sw.) Par.; *Hypnum Patientiae Americanum* R. & C. Collected Sanford, Fla.

Mr. Edward B. Chamberlain, 1830 Jefferson Place, Washington, D. C. *Hypnum uncinatum* Hedw. c.fr. Collected Cumberland, Me.

Mr. Charles C. Plitt, 1706 Hanover street, Baltimore, Md. *Catharinea angustata* Brid. c.fr. Collected near Baltimore.

Rev. S. M. Newman, corner 10th and G streets, Washington, D. C. *Hypnum ochraceum* Turn. st.; *Dicranum undulatum* Ehrh. c.fr. Collected Northwestern Adirondacks, N. Y.

Mr. J. Franklin Collins, 468 Hope street, Providence, R. I. *Aulacomnium androgynum* (L.) Schw., gemmae. Collected Cutler, Me.

Miss Alice L. Crockett, Camden, Me. *Collema flaccidum* Ach.; *Leptogium myochroum* (Ehrh., Shaer.) Tuckerm. Collected Camden, Me.

Mrs. Carolyn W. Harris, 125 St. Mark's avenue, Brooklyn, N. Y. *Leptogium tremelloides* (L. fil.) Fr. Collected Pyramid Lake, Essex Co., N. Y.

VOLUME VII.   NUMBER 4

JULY , 1904

THE BRYOLOGIST

AN ILLUSTRATED BIMONTHLY DEVOTED TO

NORTH AMERICAN MOSSES

HEPATICS AND LICHENS

EDITORS:

ABEL JOEL GROUT and ANNIE MORRILL SMITH

CONTENTS

Entered at the Post Office at Brooklyn, N. Y., April 2, 1900, as second class ot mail
matter, under Act of March 3, 1879.

Published by the Editors, 78 Orange St., Brooklyn, N. Y., U. S. A.

PRESS OF MC BRIDE & STERN, 97-99 CLIFF STREET. NEW YORK

THE BRYOLOGIST

BIMONTHLY JOURNAL DEVOTED TO THE STUDY OF NORTH AMERICAN MOSSES
HEPATICS AND LICHENS.

ALSO OFFICIAL ORGAN
OF THE SULLIVANT MOSS CHAPTER OF THE AGASSIZ ASSOCIATION.

Subscription Price, $1.00 a year. 20c. a copy. Four issues 1898, 35c. Four issues 1899, 35c.
Together, eight issues, 50c. Four issues 1900, 50c. Four issues 1901, 50c. Four Vols. $1.50
Six issues 1902, $1.00. Six issues 1903, $1.00.

Short articles and notes on mosses solicited from all students of the mosses. Address manuscript to A. J. Grout, Boys' High School, Brooklyn, N. Y. Address all inquiries and subscriptions to Mrs. Annie Morrill Smith, 78 Orange Street, Brooklyn, N. Y. For advertising space address Mrs. Smith. Check, except N. Y. City, MUST contain 10 cents extra for Clearing House charges.

Copyrighted 1904, by Annie Morrill Smith.

THE SULLIVANT MOSS CHAPTER.

President, Prof. J. M. Holzinger, Winona, Minn. Vice-President, Mrs. C. W. Harris,
125 St. Marks Avenue, Brooklyn, N. Y. Secretary, Miss Mary F. Miller. 1109 M Street,
Washington, D. C. Treasurer, Mrs. Smith, 78 Orange Street, Brooklyn, N. Y.
Dues $1.10 a year, this includes a subscription to THE BRYOLOGIST.
All interested in the study of Mosses, Hepatics, and Lichens by correspondence are
invited to join. Send dues direct to the Treasurer. For further information address the
Secretary.

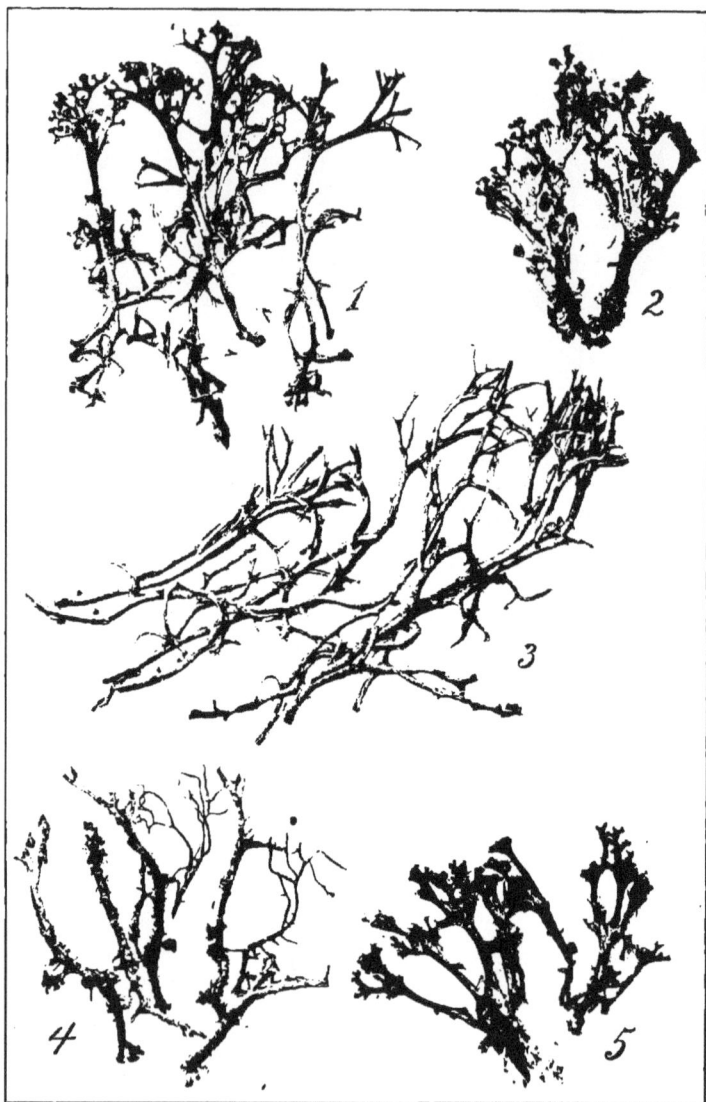

PLATE VII. Fig. 1. *Cladonia furcata* var. *racemosa*. Fig. 2. Var. *Finkii*. Fig. 3. Var. *primata*. Fig. 4. Var. *scabriuscula*. Fig. 5. Var. *paradoxa*. ✕ 1.

Vol. VII.   July, 1904.   No. 4.

FURTHER NOTES ON CLADONIAS.—III.
Cladonia furcata and Cladonia crispata.

BRUCE FINK.

After *Cladonia fimbriata*, perhaps the assemblage of lichens included in the two species in the above title are as troublesome as any. However, Tuckerman, in his treatment of the various forms of these two species, came much nearer to a correct solution than he did with regard to *C. fimbriata*, the species treated in the last paper of this series (BRYOLOGIST, 7:2. 1904). Indeed, though *C. furcata var. crispata* of Tuckerman's "Synopsis" has *seemed* difficult to trace, and though *C. furcata var. pungens* has seemed hardly to belong with the species, yet the disposition has been as a whole fairly satisfactory. Wainio has seen fit to remove the latter variety from the species, placing it with *Cladonia rangiformis*, and this appears surely to be an improvement. The former variety Wainio has also removed from the species under the name, *Cladonia crispata*. This species as viewed by Wainio seems to be well represented in Europe, where there are quite a number of varieties. However, in America, we have as yet only two of the varieties, and there is room for doubt as to whether, for our purpose, it is best to consider these forms as distinct from *Cladonia furcata*. Indeed, our *Cladonia crispata* var. *infundibulifera* seems very near to *Cladonia furcata* var. *paradoxa*, and further study is necessary to decide whether Wainio's view is the best one. But, though there may be some doubt as to best disposition of the puzzling *Cladonia crispata*, the study of the Minnesota *Cladonias* has brought to light one new variety within the two species, two others not previously known in North America, and still another known only through a single specimen collected many years ago by Tuckerman.

Regarding the illustrations, we are fortunate enough this time to be able to give them all from material that has been examined by Dr. Wainio. As to the American distribution of the varieties, it will be readily seen that with the exception of two of the first species, little is definitely known. A large amount of material in various herbaria I have not been able to examine, and a study of this would add greatly to the distribution. For the sake of the information to be gained, I should be willing to examine the material in any herbarium under either of the specific names given in our title above, and should likewise gladly examine material recorded as *Cladonia fimbriata*, the species treated in the last paper of the series. However, this would still leave untouched any material belonging to these three species, but placed elsewhere in various herbaria. With this much of preliminary statement, we may now consider the various forms of the two species.

The May BRYOLOGIST was issued May 2nd, 1904.

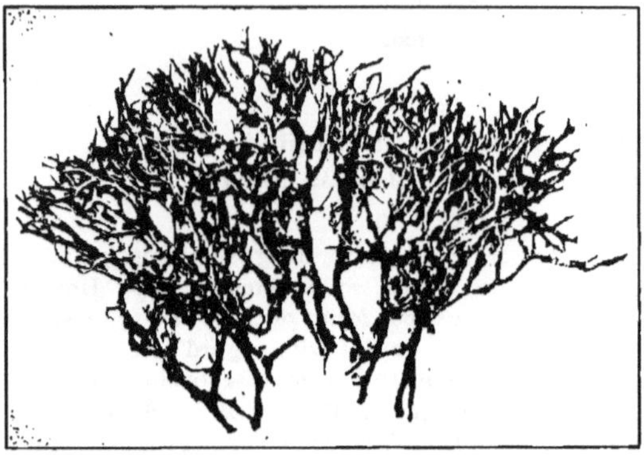

Fig. 1. *Cladonia furcata* × 1.

CLADONIA FUR-
CATA (Huds.)
Schrad. Spicil. Fl.
Germ. 107. 1794.
Fig. 1.

Primary thal-
lus usually dis-
appearing, but
when present com-
posed of medium
sized squamules,
which are cren-
ately or irregularly
lobed or rarely
subentire. .2-5
mm. long and
wide, ascending
or flat, scattered or clustered, sea-green varying toward brownish or
whitish above and white below, the cortical layer continuous. Podetia
arising from the surface of the evanescent squamules, the lower portion dying away and the apical growth continuing, 15-85 mm. long and .7-2
mm. in diameter, cylindrical or subcylindrical, rarely scyphiform, dichotomously or more or less irregularly or radiately branched, erect or rarely
decumbent or even prostrate, rarely somewhat sorediate, the cortex continuous, subcontinuous or more or less dispersed. smooth or rarely subrugose,
sometimes more or less squamulose. sea-green varying toward whitish or
brownish. the branches suberect, divaricate or recurved, the axils somewhat
dilated and frequently perforated, the apices suberect or recurved, slender
and delicate. Apothecia small, .5-1.5 mm. in diameter, irregularly or
cymosely disposed at the apices of the branches, immarginate, sometimes
lobate or reniform, convex and rarely perforate at the centre, brown varying
toward brick-red or a paler color. Hypothecium pale. Hymenium brownish above and cloudy below. Paraphyses commonly thickened and brownish
toward the apex. Asci clavate or cylindrico-clavate, the apical wall not
always thickened.

Widely distributed in North America in one form or another, most of the
material, however, being assignable to one of the varieties below. Plants
not belonging to the varieties are frequent enough where I have collected in
parts of Iowa and Minnesota. The plant used for illustration and examined by Dr. Wainio was collected at Fayette, Iowa. Macoun's "Canadian
Lichens," No. 53. belongs here and not in the variety below. Cosmopolitan
also in its foreign distribution. The form figured was sent out as the species
in my "Iowa Lichens" of 1894-5, and may be found in a large number of
American and European herbaria.

CLADONIA FURCATA (Huds.) Schrad. var. RACEMOSA (Hoffm.) Flk. Clad. Comm. 152, 1828. Plate VII. Fig. 1.

Podetia of the full length of the species or even reaching 150 mm., dichotomously or in part subradiately branched especially toward the apex, cylindrical or subcyclindrical, slightly or considerably thickened at the frequently cleft or plainly open axils, the sides closed or more or less open in places, without squamules or squamulous toward the base, cortex subcontinuous or more or less areolate, color as above or perhaps more inclined toward variegated conditions. Apothecia quite commonly present and on corymbose or cymose branches.

Dr. Wainio says, "In America septentrionali haud est rara," basing his statement upon the distribution given in Tuckerman's "Synopsis," and citing Nos. 32 and 33 of Tuckerman's "Lichenes Americani septentrionalis," these two numbers coming from the White Mountains. Though the view of the two men regarding the variety is not quite the same, we may perhaps accept Tuckerman's statement that the variety is general in its northern distribution, also "probably occurring in the southern states, at least in the mountains." However, the variety is not common where I have collected in the west, and of all the material sent Dr. Wainio, he places here only the plant collected in 1848 by C. C. Parry (Fink, B. Proc. Iowa Acad. Sci. 2:137. 1895), and of this Wainio says, "C. furcata v. racemosa in v. Finkii transiens." Calkins' "North American Lichens," No. 93, from Tennessee, and "Lichenes Boreali-Americani," Nos. 61 and 184, from Virginia and New Hampshire, are good representatives of the variety, though the varietal characters are by no means well shown in all the specimens sent out under these numbers. Also a plant from Delaware, collected by A. Commons belongs here, but also approaches var. Finkii. I have collected the variety at Cambridge, Massachusetts, and have it from the White Mountains, collected by W. G. Farlow. Thus the typical specimens seem to come from the east, and there is much doubt about the western distribution. Finally the variety is often confused with plants of other species such as Cladonia rangiferina and the two allied species, and also Cladonia amaurocraea. Known in all of the grand divisions except Africa.

CLADONIA FURCATA (Huds.) Schrad. var. FINKII Wainio Minn. Bot. Stud. 3:217. 1903. Plate VII. Fig 2.

Podetia rather stout and from 15 to 75 mm. in length and 1 to 3 mm. in diameter, scyphiform and frequently two or three ranked, cortex subcontinuous, usually more or less squamulose even toward the top, whitish seagreen or slightly olivaceous, ultimate branches sometimes quite similar to those of var. racemosa but more irregular, quite commonly fruited. Cups irregular and sometimes perforate, commonly proliferate and the cups of the upper ranks not often developed. Apothecia quite commonly present and of the usual size, color and form, occurring frequently clustered and also rarely perforate.

Dr. Wainio has failed to send a description, and I have been obliged to supply the above. However, Wainio states as follows, "Scyphifera te

—56—

analoga f. *paradoxae* Wainio, in quam transit, et e var. *racemosa* est evoluta, et in colore congruens."

Examined by Wainio from several localities along or near the northern boundary of Minnesota (Minn. Bot. Stud. 3:217. 1903). Also Nos. 767 and 914 in Minn. Bot. 2:264. 1899, are intermediate between this variety and the last, and are very similar to the C. C. Parry plant examined by Wainio. Not known elsewhere.

CLADONIA FURCATA (Huds.) Schrad. var. PINNATA (Flk.) Wainio Mon. Clad. Univ. 1:332. 1887. Plate VII. Fig. 3.

Podetia rather long and stout, the branching and condition of the axils much as in the next, but usually even less conspicuously branched toward the apex where the sterile branches are more commonly narrowly subulate, rarely decorticate in part, more or less squamulose even toward the top with incised or lobate-crenate squamules, more commonly whitish or sea-green, not isidioid or sorediate as in the next. Apothecia as usual, but not common except in the subvariety *truncata* Flk. Clad. Comm. 145. 1828, to which Dr. Wainio referred my specimen from Minnesota, and which has more obtuse apices of the ultimate branches.

Besides my Minnesota form referred to the subvariety, Dr. Wainio examined a plant from Chester, South Carolina, by H. A. Green, which he referred to the variety and which is figured. Also credited by Wainio from Great Bear Lake, Vancouver Island, New York and Mexico. Though the plant is little known, the widely separate localities would seem to indicate a general North American distribution. Known in all the grand divisions except Africa.

CLADONIA FURCATA (Huds.) Schrad. var. SCABRIUSCULA (Del.) Wainio Mon. Clad. Univ. 1:339. 1887. Plate VII. Fig. 4.

Podetia rather straight and sparingly dichotomously branched (especially toward the apex), apices usually subulate, more or less isidioid or sorediate and also commonly sparingly or even densely squamulose, frequently also with the cortex more or less broken or even partly decorticate, usually whitish to sea-green. Apothecia apparently rare, and scarcely ever present on our American specimens.

Dr. Wainio has determined this for me from Minnesota and Iowa, the Iowa plant having been distributed by me in 1894-5 in the "Lichens of Iowa" as var. *racemosa*, after comparing with material in the Tuckerman herbarium at Harvard. Of this material Wainio says, " *Cladonia furcata* X *C. furcata v. scabriuscula*," but I think that my plants distributed all show the varietal characters clearly. Numbers 47 and 51 of Macoun's "Canadian Lichens" I find to be this variety, and I have received it from Newfoundland through A. E. Waghorne. Wainio credits the variety from New Bedford, Massachusetts, under the subvarietal name *C. furcata* var. *scabriuscula*, forma *farinacea* Wainio Mon. Clad. Univ. 1:339. 1887, but I have not seen this form. Not known elsewhere in North America. Known in all the grand divisions except Africa.

CLADONIA FURCATA (Huds.) Schrad. var. PALAMAEA (Ach.) Wainio Mon. Clad. Univ. 1:347. 1887.

Podetia usually dichotomously or subradiately-ramose branched, cylindrical and frequently somewhat thickened toward the axils, sometimes squamulose toward the base, chestnut-colored or brick-red, or olivaceous varying toward sea-green toward the base. Apothecia as usual.

Dr. Wainio says, "Verisimiliter etiam in America septentrionali distributa est (var. *subulata* Tuck., Syn. North Am. p. 248 pr. p.)," thus leaving some doubt as to whether this form of the species is North American. However, Wainio has determined what he regards a subvariety from Minnesota. This is given next below, and I shall regard it as a distinct variety till I have an opportunity of examining specimens of the above, or am able to see more of resemblance than appears in the two descriptions.

CLADONIA FURCATA (Huds.) Schrad. var. PARADOXA (Wainio) Fink Minn. Bot. Stud. 3:217. 1903. Plate VII, Fig. 5.

Podetia rather short, 10-40 mm., and .7-1.5 mm. in diameter, scyphiform, brownish or olivaceous-brown above and sometimes sea-green toward the frequently sparsely squamulose base. Cups 2-3 mm. in diameter and quite abruptly dilated, the cavities perforate or subcribose, irregularly sublacerate-proliferate, the proliferations forming two or three ranks and even the highest rank commonly scyphiform. Apothecia not conspicuously clustered, often perforate, or lobate, brown, convex, immarginate, quite common at least in ours.

Determined by Dr. Wainio from northern Minnesota, where frequent on old wood or thin earth. Otherwise only known in Europe. Seems very near to the first variety of *Cladonia crispata*, below, from which it may be distinguished readily by the difference in color and some features of the primary squamules, which are more commonly persistent in *Cladonia crispata*.

CLADONIA CRISPATA (Ach.) Flt. Merkw. Flecht. Hirschb. 4. 1839.

Primary thallus persistent or finally dying, composed of middling sized digitate-laciniate or crenate squamules, which are 1-4 mm. long and wide, ascending, flat or involute, scattered or rarely clustered and forming a compact crust, lighter or darker sea-green or even olivaceous-brown above and white or brownish below or even reddish below toward the base, the cortex continuous. Podetia arising from the surface of the squamules, the base often dying away and continuing to grow above, 10 to 75 or possibly 100 mm. in length and .5 to 5 mm. in diameter, subcylindrical or irregularly turgescent or even trumpet-shaped, radially or sympodially branched, the branches suberect or spreading, the axils commonly somewhat dilate-open, the cortex subcontinuous or dispersed-areolate and the areoles frequently more or less raised, sometimes more or less squamulose, sea-green or variously whitish, reddish, brownish or olivaceous, most commonly scyphiform, the ultimate branches also scyphiform, or obtuse or subulate. Cups abruptly dilated and frequently perforate, borne at the apices of the branches, repeatedly proliferate at the margin, the species probably not commonly more than one ranked. Apothecia small, .5-.7 in diameter, at the apices of short branches or at the ends of the proliferations of the cups, subsolitary or subcorymbosely aggregated, immarginate or with thin margin, flat or convex, brown or rarely brick-red. Hypothecium pale. Hymenium pale or pale-brownish below and brownish above. Paraphyses commonly simple, thickened but usually pale at the apex. Asci cylindrico-clavate, the apical wall thickened.

The same in part as *Cladonia furcata* var. *crispata* of Tuckerman's "Synopsis," and Dr. Wainio credits it from Great Slave Lake, British Columbia, Rocky Mountains, Wisconsin, California and Massachusetts, and has determined it for me from Minnesota. This gives a general distribution throughout northern United States and northward. Known in all the grand divisions except Africa.

Fig. 2. *Cladonia crispata var. infundibulifera* × 1.

CLADONIA CRISPATA (Ach.) Flt.var. INFUNDIBULIFERA (Schaer.) Wainio Mon. Clad. Univ. 1:382. 1887. Fig. 2.

Podetia rather longer and stouter, sometimes squamulose toward the base, scyphiform, sometimes two or three ranked. Cups perforate, commonly abruptly dilated, 3-6 mm. wide, regular or finally oblique, radiate or proliferate. Apothecia on the proliferations or on short pedicels on the margins of the cups.

Dr. Wainio has determined this variety for me from northern Minnesota, where I have made three collections, and refers number 31 of Tuckerman's "Lichenes Americani septentrionalis" here. Not known elsewhere from North America. Tuckerman's number 31 was collected in the White Mountains. Known also in Europe.

CLADONIA CRISPATA (Ach.) Flt. var. SUBCRISPATA (Nyl.) Wainio Mon. Clad. Univ. 1:385. 1887.

Podetia about 45 mm. long, scyphiform, sparsely squamulose, suberect or recurved, sea-green. Cups with the cavity usually closed or rarely cribrose or perforate, quite regular in form, sometimes irregularly perforate or radiate. Apothecia on corymbose branches or on the margins of the cups.

Dr. Wainio regards this a strictly North American variety, his note on distribution being as follows, "In partibus Britannicis Americae septentrionalis," but the form is wholly unknown to me.

The material examined is as before from my own collecting, or from my herbarium or the very full one at the University of Minnesota, and was collected by Farlow, Seymour, Miss Cummings, Willey, Calkins, Eckfeldt, Waghorne, Tuckerman, Parry, Green, Commons and others. I have again found the European material sent by L. Scriba very helpful for comparison, but the illustrations are this time all from American plants, examined by Dr. Wainio.   Grinnell, Iowa.

FURTHER NOTES ON SEMATOPHYLLUM.

ELIZABETH G. BRITTON.

In the BRYOLOGIST for January, 1903, I stated that two attempts had been made to see the types of *Leskea recurvans* and *Leskea squarrosa* Michx. but without success. They were not to be found at the Jardin des Plantes nor in the Herbarium of M. Drake del Castillo. Miss Vail, on her visit to Paris last year, obtained information that they were in the Herbarium of M. Ferdinand Camus, who very kindly offered to loan them to me, so that at last I have been able to compare them with North American specimens distributed in Exsiccatae and with illustrations. Both the type specimens came from Richard's herbarium and the notes accompanying them are in his handwriting and are almost identical with the descriptions published by Michaux (Flor. bor. Am. 2:311. 312. 1803) with a few omissions and transposals. These descriptions have been compared with the types.

LESKEA RECURVANS Rich. The type specimens of " *Leskea? recurvans* " are labelled with a question mark as here indicated. There are three tufts in good condition showing an abundance of capsules both with and without the lids. The old capsules are contracted below the mouth when dry as in Plate I, Fig. 12, BRYOLOGIST, 6:1, 1903, and all are more or less curved and unequal. The alar cells of the leaves are more inflated than figured by Sullivant in the Icones, Plate 111, Fig. 7, as much so as Icones Supplement, Plate 69, Fig. 6, and this confirms Prof. D. C. Eaton's and C. F. Austin's criticisms of these plates.

The seta varies in length from "6-8 lines," and the capsules also differ considerably in size (1-2 mm.) and degree of development. The lid also varies in length from the rostrate-apiculate shape figured by Sullivant, Icones Musci, t, 111, Figs 11-12, to even longer and more curved forms as in Sull. Musci Alleghaniense No. 18, and Sull. & Lesq. Musci Bor. Am. No. 301b. (These specimens were distributed as *Hypnum recurvans* var.) *Leskea squarrosa* was cited as a synonym in No. 18 Musci Alleghaniense. Considering that one hundred years have elapsed since the specimens were collected their macroscopic appearance has not greatly changed. They still retain their lustre and color and are undoubtedly what has been generally known as *Hypnum recurvans* Schwaegr.

LESKEA SQUARROSA Richard. These specimens were old and in poor condition when collected. The label states that they grew " *in humosis humidis* " and were collected by *Beauvois*. They are mixed with hepatics which give them a darker green color and only a few capsules are preserved; these are old and deoperculate. The description states that the lid was not seen. The pedicel is described as " subunciales," apparently they are all shorter than in the type of *Leskea recurvans*, only about one-half inch long. A portion of *Hypnum recurvans*, No. 301. Sull. & Lesqx. Musci Bor. Am. in the Columbia Herbarium collected agrees exactly with these type specimens of *Leskea squarrosa*, and indicate that the differences are due to the greater moisture of the locality where they grew.

As previously stated in the BRYOLOGIST, specimens of *Hypnum recurvans* have been found growing on wet cushions of *Leucobryum*, where the leaves were scarcely recurved and distinctly "*squarrose*." It is evident that *Leskea squarrosa* represents a form of *Leskea recurvans* with more spreading leaves, just as *Hypnum laxepatulum* is a similar form of *Semato phyllum delicatulum*. These differences are not of specific importance, for the leaves may be flat and spreading on the younger branches and strongly recurved on the older ones of the same tuft, as seen in the type specimens. The leaves are long-pointed and sharply serrate at apex and the alar cells yellow and vesicular as in all the forms and varieties of this species.

Sematophyllum recurvans (Rich.) E. G. B. BRYOLOGIST **5**:65. 1902.

Leskea recurvans (Rich.) Mx. Fl. Bor. Am. **2**:311. 1803.

Hypnum recurvans Beauv. Prod. Aetheog. 73. 1803.

Raphidostegium recurvans Jaeg. & Sauerb. Adumb. 400. 1877.

Plants in glossy yellowish-green mats, pulvinate, with irregularly pinnate, creeping stems on the circumference, branches simple or divided, up to 10 or 15 mm. long, erect or decumbent, crowded with circinate leaves, recurved on both sides of the stem, with few small paraphyllia, leaves 1 mm. long, acuminate, serrate above the middle, the margins slightly revolute, basal cells yellow, porose, alar cells larger, rounded and inflated, upper linear-vermicular, perichaetial leaves longer, more acuminate and more sharply serrate at apex. Dioicous, the antheridia in small, leafy axillary buds, antheridia with few paraphyses. Seta glossy, red-brown. 10-30 mm. long, capsule contracted below the mouth when dry; lid conic-rostrate, the beak varying from once to twice the length of the conical part of the lid; walls much thickened and brown, neck stomatose; peristome double, yellow, paler than the walls; teeth striate at base, papillose at apex, the inner face deeply lamellate; endostome shorter with one or two slender papillose cilia; spores smooth, .010–.015 mm. maturing in summer and autumn.

Type locality: Mts. of Carolina on wet rotten logs, Michaux.

Type specimen preserved in Herb. Richard, the property of M. Ferdinand Camus, at Paris.

Exsiccatae: Drummond N. Am. Mosses No. 196 in part 1828. Sull. Musci All. No. 18. 1848. Sull. & Lesqx. Musci Bor. Am. No. 301–301b. 1856. Nos. 446 and 447 in part 1865. Austin Musci App. Nos. 340–341. 1870–7. Macoun Canadian Mosses No. 229; Small Mosses S. U. S. No. 14.

Grout N. Am. Musci Pleu. No. 1, agrees very well with the type of *Leska recurvans*.

Distribution: N. B., Nfd., Quebec, Ont., to Manitoba. Eastern U. S. from Maine to N. C. Central States from Minnesota to Missouri.

VAR. SQUARROSA (Brid.) Britton. n. var.

Leskea squarrosa Rich. Michx. Flor. Bor. Am. **2**:312. 1803.

Plants in thin mats. Stems creeping, bright yellow and glossy or darker green, the leaves crowded, more flattened and spreading on the stems and less strongly recurved, though variable on the same plant; pedicels shorter often only 10 mm. long; capsule smaller, 1 mm.

Growing in moist humus or on damp cushions of *Sphagnum* or *Leuco-bryum* often mixed with other mosses and hepatics.

Type locality: Carolina, Beauvois, Michaux.

Type specimen preserved in Herb. Richard, property of M. Ferdinand Camus.

Exsiccatae: No. 62 Ohio, W. S. Sullivant 1842. Sull. Musci All. No. 17,1848. Sull. & Lesqx. Musci Bor. Am. No. 301pp. 1856. Nos. 446pp. and 447. 1865.

No. 62, was labelled "*Hypnum cupressiforme* var." of W. S. Sullivant's Ohio Mosses distributed in 1842 with written labels, is part of a set which seems to have been the first attempt to prepare *exsiccatae* of North American mosses that Sullivant made. They were put up in small square volumes, one set of which is still preserved in its original form in the Herbarium of the Academy of Natural Sciences in Philadelphia.

Distribution: Cedar Swamp, Hackensack, N. J., Torrey; Kingwood, N. J., Best; Westville, Conn., Chatterton; Tenn., Kearney; Manitoba, Macoun.

VAR. COMPACTA (Aust.) Ms. in Herb.

Plants forming dense glossy yellow tufts matted with brown radicles. Stems erect, 2-3 cm. high; branches short and crowded with appressed leaves not flattened, pedicels often longer than the type, 15-30 mm. long.

Habitat in rich humus in wet soil or swamps. Otter Pond, N. J., Austin: Cedar swamps, N. J., and Catskill Mts., N. Y.; Tibbs Run, W. Va., C. F. Millspaugh.   New York Botanical Garden.

MOUNTING MOSSES.

B. D. GILBERT.

During last year I read and tested all the ways of mounting mosses that had been described in the BRYOLOGIST, but not one of them proved satisfactory. I wanted the mosses in such shape that they could be lifted with a pair of forceps and examined. This barred out the plan of glueing the specimens to a piece of paper or cardboard. The alternative of course was to put the specimens into pockets. But what was to be done with the specimens afterward, so that any particular species could be turned to in half a minute or less? I struggled with this problem at intervals of several months, until at last it worked out in the manner which is here described.

I bought a thin white linen paper of medium size, i. e., 17 x 22 inches, had this cut in two the short way, then cut in two again, and a third time cut in two. This left sheets 5½ x 8½ inches. I then folded each sheet the long way so that the flap was about 1¼ inches wide, turning the ends backward about a half to five-eighths inch. All of these folds can be guessed at with sufficient accuracy. The piece of the flap folded over at each end was then cut out, leaving the flap perfectly loose and free, while at the same time it was impossible for the moss enclosed to slip out. On the flap was written the name, locality and collector, the specimen with its label, if it had one, was enclosed and then the pocket was ready to be put in place.

It was my aim to use the same size and quality of sheets for mosses as was used in my general herbarium viz. 11 x 17 inches. Eight of the pockets will lie on a sheet of this size and leave considerable room between. I begin by placing two pockets side by side at the bottom of the page and stretching a strip one inch wide of the same stiff paper as the sheet itself across the page high enough up to let the flap of pocket fold over it at the joint or crease. This strap is held in place by half inch strips of commercial gummed paper at each end and one in the middle. This is repeated until all the eight pockets are in place, and then you have a sheet which can be handled as carelessly as you please without disordering your specimens in the least; while on opening the sheet you can see at a glance what it contains, select your pocket and remove it with perfect ease. On the lower left hand corner outside of the sheet can be inscribed the name of the genus which it contains, or the letter with which several genera enclosed begins and the sheets placed in alphabetical order.

I have as yet only about one hundred and fifty species, but should arrange them the same if I had one or two thousand species. The facility of reference more than makes up for lack of scientific arrangement which must lie in one's head rather than in the sheets. Of course the Hypneae are kept in a cover by themselves and their different genera or subgenera arranged in similar alphabetical order.

Several genera of mosses like Antitrichia, Fontinalis, Hylocomium, Neckera, Sphagnum and the others require pockets of double size. Four of these will go on a page if slipped under the straps endwise without reference to the flaps. I find that an average of ten to twenty sheets containing both small and large size pockets will hold about six pockets to the sheet. Consequently it will be seen that it takes comparatively few sheets to hold quite a respectable collection of mosses, especially as two or three duplicate pockets of the same species can be slipped into one loop together. The trouble of mounting is very small while the ease of handling and selecting is very great. Clayville, New York.

MUSCI BOREALI-AMERICANI BY PROF. J. M. HOLZINGER.

A. J. Grout.

Fascicles II and III of Prof. Holzinger's Musci Acrocarpi Boreali-Americani were issued in rapid succession; both contain numerous interesting species. Fascicle II contains among other interesting species: *Polytrichum Smithiae* Grout, *Anomobryum filiforme Americanum* R. & C., *Fissidens decipiens Winonensis* R. & C., *Coscinodon Raui* (Aust.) L. & J., *C. Wrightii* (Aust.) Sull., *Bryum Duvalii lato-decurrens* C. M. & Kindb., *B. Sawyeri* R. & C., *Catherinea mollis Holzinger* and *Dicranum fragilifolium* Lindb. Fascicle III contains the following: *Syrrhopodon Floridanus* Sull., *Orthotrichum cupulatum Porteri* Vent., *Schistidium alpicola rivulare* (Brid.) Wahlenb., *Tetraplodon Australis* Sull. & Lesq., *Barbula Raui* Aust., *Amphidium Mougeotii* (B. & S.) W. P. Sch. and *Bruchia curviseta* L. & C.

No student of North American mosses should feel satisfied without Prof. Holzinger's exsiccati.

PART II. MOSSES WITH HAND-LENS AND MICROSCOPE.

J. FRANKLIN COLLINS.

The second part of Dr. Grout's new book is a pamphlet of about the same size as, and a continuation of, Part I. It completes the family *Dicranaceae* with keys and descriptions of thirty-two species and several varieties, and includes the genera *Ceratodon, Sælania, Trematodon, Dicranella, Blindia, Dicranodotium, Dicranum* and *Leucobryum.* The *Grimmiaceæ* comprises the genera *Hedwigia* and *Ptychomitrium,* each with a single species, *Grimmia,* with ten species and a few varieties, and *Rhacomitrium* with six species. *Ephemeraceæ* includes the small genera *Nanomitrium, Ephemerum, Acaulon* and *Physcomitrella.* The remaining two-fifths of Part II comprises the family *Tortulaceæ.* Under this topic about thirty species and several varieties of the genera *Phascum, Astomum, Weisia, Gymnostomum, Didymodon, Barbula, Trichostomum, Tortella, Pottia, Pterygoneurum, Aloina, Desmatodon* and *Tortula* are described. Both the text and the copious illustrations are of the same high grade as in Part I. (See THE BRYOLOIST, 6:104.)

In botanical works which particularly aim to instruct, interest and maintain the interest of the novice illustrations have long played an important part. Unfortunately for the beginner these illustrations are sometimes far from being above criticism as regards accuracy, and at times are positively misleading, as many an older student may personally recollect. In reproducing so many of his illustrations from the Bryologia Europaea and Sullivant's Icones—the two great illustrated bryological works of Europe and America respectively—Dr. Grout has left few, if any, openings for criticism of this nature.

It is undoubtedly impossible to mechanically reproduce a plate and have it equal to the orginal. As a test of the quality of the plates in Part II the reviewer has made a careful comparison with the original of each plate illustrating *Dicranaceæ* which has been reproduced from the Bryologia, the Icones or Limpricht. When placed side by side a difference is at once detected, yet a difference which in no way involves the question of inaccuracy or carelessness. We see prints from carefully made, full-sized, photographically reproduced plates of the originals. For all ordinary working purposes the beginner (and most others) will find them as useful as the originals and far less expensive.

Some of the moss names will not be wholly approved by certain bryologists but as this involves largely a question of personal conviction or opinion, and in no way impairs the value of the work, it would be out of place to discuss them here. The accented technical names will be greatly appreciated by most beginners. No student of mosses within the Gray's Manual region should be without " Mosses with Hand-Lens and Microscope."

Providence, R. I.

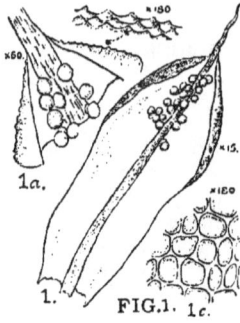

1a.

1.

FIG.1. 1c.

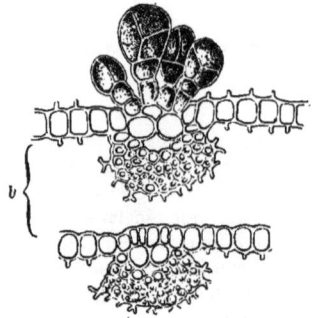

b

FIG. 2.

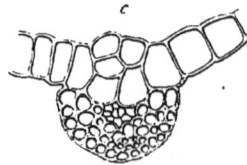

c

FIG. 3.

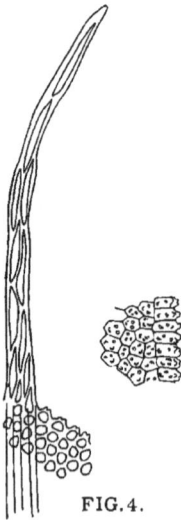

FIG. 4.

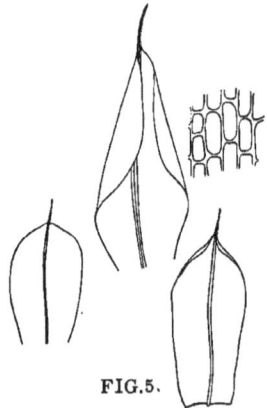

FIG.5.

PLATE VIII.   *Tortula.*

TORTULA PAGORUM (MILDE.) DE NOT. IN GEORGIA.

A. J. GROUT.

In working over my collection of Tortulaceae as a preparation for writing up the family for "Mosses with Hand-Lens and Microscope" I found that Dr. John K. Smalls' "Mosses of the Southern United States," No. 39, on Ulmus, Atlanta, Georgia, July 1, 1895, labelled *Barbula papillosa* (Muell.) Wils. was not that species at all, and it was evidently no species credited to North America.

The plants are short, 3-5 mm., stout, very dark or brownish green leaves when dry appressed and somewhat contorted, when moist erect-spreading, lingulate to obovate or fiddle-shaped, margins plane below entire, occasionally incurved at the middle, sometimes emarginate according to Limpricht, but in our specimens not emarginate and with the upper margins incurved as in *T. papillosa* as is shown in the illustrations, Plate VIII. Leaves strongly papillose above with papillae of various shapes and degrees of complexity as shown in the illustration; lower leaf cells much larger, rectangular, nearly hyaline; costa excurrent, smooth at back. In the axils of the upper leaves are numerous leaf-like, ecostate broodbodies, having strongly papillose cells and a hyaline apiculus.

Easily known from *T. papillosa* which has simple papilla on the leaf cells, larger forked papillae on the back of the costa and numerous clavate to subglobose, multicellular broodbodies clustered on the upper surface of the costa. The illustrations will serve to make the distinctions clear. *T. papillosa* is probably common on the bark of trees, and I expect that it will be frequently reported now that its determination is made easy. *T. pagorum* is likely to be found much farther north and herbarium specimens labelled *T. papillosa* should be carefully examined for it. Both species rely on their broodbodies for reproduction. The fruit of *T. pagorum* has never been collected, while the only fruit of *T. papillosa* known to science came from Australia and New Zealand.

Plate VIII, Fig. 1. *Tortula papillosa*, 1, Leaf showing brood bodies on costa, 1a, apex of leaf showing papillae and brood bodies. After Dixon and Jameson.

Fig. 2. Cross sections of leaf showing brood bodies and papillae. After Limpricht.

Fig. 3. *Tortula pagorum*, brood bodies highly magnified, "c", cross section of costa much less magnified. After Limpricht.

Fig. 4. *T. pagorum*, apex of leaf and marginal papillae.

Fig. 5. Leaves and cells of same. Figs. 4 and 5 after drawings by Miss Thayer.

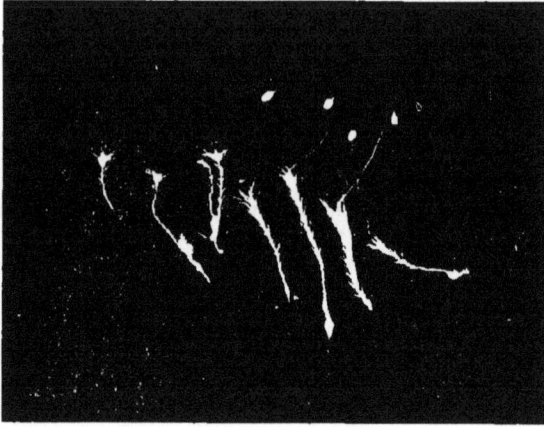

Webera Lachenaudi Card & Ther. n. sp. Plate IX.

A new Webera has been collected recently in our locality, *Webera Lachenaudi* Card. & Ther., named after M. Georges Lachenaud, of Limoges, France. The following are M. Cardot's notes on it. (Translated.) " *Webera Lachenaudi* C. & T. (species nova), near to *Webera lutescens* Limpr. It differs in its larger stems, its strongly dentate leaves, its more chlorophyllose tissue and its capsule more urceolate (not pyriform)."

It is found on wet sand and clay banks in the woods south of the north bay of Lake Union. Fruit and male inflorescence shown in the illustration.

JOHN W. BAILEY, M. D.,

Seattle, Wash.

THE HULST BOTANICAL CLUB OF BROOKLYN, N. Y.

On Tuesday afternoon. April 13th, 1904, the following persons met at Dr. A. J. Grout's home to organize a botanical club. Mesdames Alice H. Shepard, Margaret H. Platt, Rebecca L. Palmer, Emma L. Kingsland. Caroline A. Creevey, Low, Ida M. Conklin, Grace Grout, Annie Morrill Smith, Carolyn W. Harris, Mr. Wood and Dr. A. J. Grout. It was decided to name the club " The Hulst Botanical Club," in memory of Dr. George D. Hulst, who for many years was president of the Department of Botany of the Brooklyn Institute of Arts and Sciences and who was probably the best informed botanist of the flora of Long Island.

The object of the Club to be the mutual benefit to its members of discussion of botanical subjects and the awakening of interest along such lines.

The special work of the Club to be the making of an exhaustive list of the flora of Long Island.

The Club is to form a Chapter of the Wild Flower Preservation Society of America. Dr A. J. Grout was appointed director of the Club; Mrs. Carolyn W. Harris, secretary.

SULLIVANT MOSS CHAPTER NOTES.

By Miss Genevieve Doran.

"The Field Day or Moss Walk," announced in the March Bryologist for all the Sullivant Moss Chapter members and their friends living within ten miles of Boston, took place on April 23d. In response twenty-five devotees of the subject assembled, nine of whom were members of the Moss Chapter. One of the members unable to be present sent a substitute. The originator of the event, Mr. Walter Gerritson, conducted the party over Prospect Hill, in Waltham. The route of the walk was arranged on the southwest slope of Prospect Hill, thus affording the best light for the collectors and also mosses, whose habitat ranged from the face of a dry rock to the shady bog. During the walk the party was afforded the pleasure of witnessing the fine view from the summit of "Little Prospect."

In all fifty different species of mosses and eleven species of lichens were collected. The lichens were identified by Miss C. M. Carr, of South Sudbury, Mass. One enthusiastic member came fifty miles to join the party, and all united in pronouncing it a profitable and enjoyable day.

The members present were: Miss E. B. Bryant, Miss C. M. Carr, Miss G. Doran, Mrs. J. J. Puffer, Mrs. L. M. Stevens, Mr. J. F. Collins, Mr. W. Gerritson, Mr. J. W. Huntington, Dr. G. G. Kennedy, and a substitute for Mrs. J. B. Clapp.

List of Lichens: Cladonia cristatella, C. pyxidata var.? Peltigera rufescens, P. polydactyla, P, canina, Parmelia caperata, P. perlata, P. perlata var.? P. saxatilis, P. conspersa.

List of Mosses: Anomodon attenuatus, Aulacomnium heterostichum, A. palustre, Bartramia pomiformis, Brachythecium plumosum, Bryum argenteum, B. nutans, Buxbaumia aphylla, Catherinea angustata, Ceratodon purpureus, Climacium Kindbergii, Dicranum flagellare, D. fulvum, D. scoparium, Dicranella heteromalla var. orthocarpa, Ditrichum pallidum, D. tortile, D. vaginans, Entodon seductrix, Funaria hygrometrica, Georgia pellucida, Grimmia apocarpa, G. conferta, Hedwigia albicans, Hylocomium proliferum, Hypnum crista-castrensis, H. Haldanianum, H. hispidulum, H. imponens, H. reptile, H. Schreberi, Pogonatum tenue. Leucobryum glaucum, Thelia hirtella, Mnium sylvaticum, Orthotrichum sordidum, Plagiothecium striatellum, P. sylvaticum, Polytrichum commune, P. juniperinum, P. Ohioense, P. piliferum, Pylaisiella intricata, Sphagnum acutifolium, S. cymbifolium, Thidium delicatulum, T. scitum, Webera sessilis, Weissia Americana, W. ulophylla.   Waltham, Mass.

Note.—The excursion on April 23d was so successful that a second was arranged for May 14th, and four members have been added to the Chapter as among the results.   A. M. S.

A NEW SPECIES OF HEPATICAE.

Taken from "The Botanical Gazette," November, 1903. From an article on "Odontoschisma Macounii and its North American Allies," by Alexander W. Evans.

ODONTOSCHISMA GIBBSIAE EVANS.

Plants yellowish green, more or less tinged with red or brown, growing in depressed mats or creeping among other bryophytes: stems sparingly and irregularly branched, 0.3 mm. in diameter, prostrate, ascending at the tips; flagella postical or terminating leafy branches; vegetable branches varying from postical to lateral; rhizoids scanty: leaves imbricated, strongly concave, increasing in size from the base of a leafy axis, orbicular, 1 mm. long, not bordered, attached by an oblique line of insertion, slightly decurrent antically and more or less dillated at postical base, arching to or beyond the middle of the axis, margin entire, apex rounded: leaf-cells 16 x 23 μ at edge, 19 μ in diameter in the middle and 23 μ at the base, with very large and occasionally confluent rounded trigones; cell cavities stellate with distinct pits; pigmentation when present limited to the lining of the cavity, not affecting the limiting membrane of the pits nor the outer part of the very thick, smooth or minutely verruculose cuticle: underleaves minute, except at the base of a branch, irregular in shape, sometimes vaguely bidentate: inflorescence unknown: gemmae and gemmiparous branches similar to those of *O. Macounii* but the latter with more loosely imbricated leaves.

On a log. British Columbia: Port Renfrew, Vancouver Island (Miss Gertrude Gibbs), the type locality, Plate XIX.

OFFERINGS.

[To Chapter Members only—for postage.]

Mr. J. Warren Huntington, Amesbury, Mass. (Four cents postage.) *Sphagnum acutifolium* Ehrh.; *S. cuspidatum* Ehrh: *S recurvans* (P. Beauv.) R. & W. var. *mucronatum* Russ.; *S. squarrosum* Pers.; *S. cymbifolium* (Ehrh.) Hedw.; *S. rubellum* Wils. Collected in Amesbury.

Mr. B. D. Gilbert, Clayville, Oneida Co., N. Y. *Funaria flavicans* Michx.

Mr. N. L. T. Nelson. 3968 Laclede Ave., St. Louis, Mo. *Gymnostomum rupestre* Sch. c.fr. Collected in Minnesota. *Weisia viridula* Brid. c.fr. Collected in Missouri.

Miss Mary R. Haughwout, Wilson College, Chambersburg, Pa. *Climacium Kindbergii* (R. & C.) Grout, c.fr.

Mr. G. K. Merrill, 564 Main street, Rockland, Me. *Evernia furfuracea* (L.) Ach. var. *Cladonia* Tuckerm. Collected in Camden.

Mrs. Carolyn W. Harris, 125 St. Mark's Avenue, Brooklyn, N. Y. *Sphaerophorus globiferus* (L.) DC. Collected by Mr. A. J. Hill, British Col.

MOSSES
WITH HAND-LENS AND MICROSCOPE

QUARTO: bound in durable pamphlet cover. Printed on the best paper by the best Printers, the Mount Pleasant Printery.

The HAND-LENS CHARACTERS are fully emphasized so that these alone are often sufficient to determine the plant and in addition the MICRO-SCOPIC CHARACTERS are treated in the same untechnical manner.

$1.00 a part, post-paid.

Checks NOT on New York City MUST be made out for 10c. extra.

PART I. 86 pp. CONTAINS A COMPLETE TEXT-BOOK
ON MOSSES TREATING OF

*Life-History, Structure, and Biology.   Illustrated Glossary, Revised and Rearranged.
Key to Families. Also a Systematic Treatment of the Families partly Through the Dicranaceae.*

PART. II. Ready May 15th. DICRANACEAE TO ENCALYPTACEAE.

88 pp. 25 full page Plates. 50 cuts in the Text.

SEND FOR SAMPLE PAGES.

Published by the Author A. J. GROUT, Ph.D, 360 Lenox Road, Brooklyn, N. Y.

Also a few sets of North American Musci Pleurocarpi. Second century just completed. $7.50 per century.   A. J. GROUT.

WANTED

BOTANICAL DRAUGHTSMAN—Biological drawing a specialty. Illustrations made for effective and truthful reproduction.

Address. P. B. WHELPLEY, Dublin, N. H.

ALL ABOUT THE WILD=FLOWERS

THERE is only one publication *about* the wild-flowers and that is *The American Botanist.* It does not publish technical articles, and uses the common names of plants whenever possible. Full of notes on the haunts, habits. fragrance, uses, and products of plants. Every plant-lover should have it. Monthly, $1.00 a year. Sample for a 2-cent stamp.

ADDRESS. **THE AMERICAN BOTANIST** ⤳ **Binghamton. N. Y.**

——————————*The*——————————
Journal of the Maine Ornithological Society

With the January, 1904, Number
Begins its Seventh Volume

A Quarterly Journal all about Maine Birds

You should have Mr. Arthur H. Norton's papers on "The Finches Found in Maine" and the series of papers on "The Warblers Found in Maine," written up by four of the members of the Society. These alone well worth the price of a year's subscription.

Subscription. Fifty Cents per Annum ⤳ ⤳ **Fifteen Cents per Copy**
Send stamps for Sample Copy to

J. MERTON SWAIN, Business Manager ⤴ ⤴ FAIRFIELD, MAINE

VOLUME VII.   NUMBER 5

☀ SEPTEMBER, 1904 ☀

THE BRYOLOGIST

AN ILLUSTRATED BIMONTHLY DEVOTED TO

NORTH AMERICAN MOSSES

HEPATICS AND LICHENS

EDITORS:

ABEL JOEL GROUT and ANNIE MORRILL SMITH

CONTENTS

Entered at the Post Office at Brooklyn, N. Y., April 2, 1900, as second class or mail
matter, under Act of March 3, 1879.

Published by the Editors, 78 Orange St., Brooklyn, N. Y., U. S. A.

PRESS OF MC BRIDE & STERN, 97-99 CLIFF STREET, NEW YORK

THE BRYOLOGIST

BIMONTHLY JOURNAL DEVOTED TO THE STUDY OF NORTH AMERICAN MOSSES
HEPATICS AND LICHENS.

ALSO OFFICIAL ORGAN
OF THE SULLIVANT MOSS CHAPTER OF THE AGASSIZ ASSOCIATION.

Subscription Price, $1.00 a year. 20c. a copy. Four issues 1898, 35c. Four issues 1899, 35c.
Together, eight issues, 50c. Four issues 1900, 50c. Four issues 1901, 50c. Four Vols. $1.50
Six issues 1902, $1.00. Six issues 1903, $1.00.

*Short articles and notes on mosses solicited from all students of the mosses. Address manu-
script to A. J. Grout, Boys' High School, Brooklyn, N. Y. Address all inquiries and subscriptions to Mrs. Annie Morrill Smith, 78 Orange Street, Brooklyn, N. Y. For advertising space address Mrs. Smith. Check, except N. Y. City, MUST contain 10 cents extra
for Clearing House charges.*

Copyrighted 1904, by Annie Morrill Smith.

THE SULLIVANT MOSS CHAPTER.

President, Prof. J. M. Holzinger, Winona, Minn. Vice-President, Mrs. C. W. Harris,
125 St. Marks Avenue, Brooklyn, N. Y. Secretary, Miss Mary F. Miller, 1109 M Street,
Washington, D. C. Treasurer, Mrs. Smith, 78 Orange Street, Brooklyn, N. Y.
Dues $1.10 a year, this includes a subscription to THE BRYOLOGIST.
All interested in the study of Mosses, Hepatics, and Lichens by correspondence are
invited to join. Send dues direct to the Treasurer. For further information address the
Secretary.

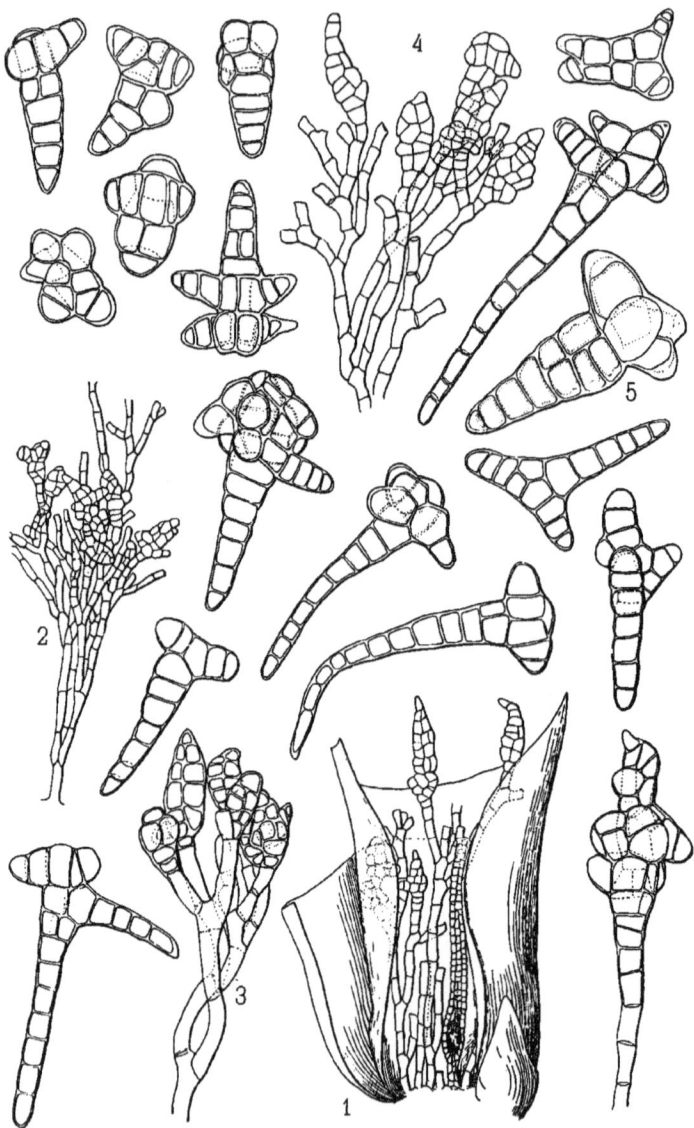

PLATE X. *Hyophila riparia.*

THE BRYOLOGIST.

Vol. VII.   September, 1904.   No. 5.

HYOPHILA—A NEW GENUS TO THE UNITED STATES.

By Elizabeth G. Britton.

Prof. Max Fleischer, whose studies on the moss-flora of Buitenzorg have made him familiar with *Hyophila Javanica* has called my attention to the fact that the much-named *Pottia riparia* Austin is really a species of *Hyophila*. I have made comparisons of the specimens and illustrations and find that he is correct.

The following generic description and synonymy have been adapted from those given by Brotherus in " Die Natürlichen Pflanzenfamilien " :

Hyophila Brid. Bryol. univ. **1** : 760. 1826.

Rottleria Brid. Bryol. univ. **1** : 105. 1826. not Willd. 1797.

Pottia section Hyophila CM. Syn. Musc. **1** : 558. 1849.

Weisia section Hyophila Mitt. Jour. Linn. Soc. **12** : 135. 1869.

Plants growing in low, dark green or brown tufts. Stems radiculose at base, branching; central strand present. Leaves crowded, inrolled and curled when dry, spreading when moist; base clasping, oblong; apex elliptic, obtuse or acute, entire or serrate: costa stout, ending in or below the apex or rarely short excurrent; cells smooth or slightly papillose. Dioicous. Perichetial leaves smaller or not differentiated, Seta exserted, slender, erect. Capsule erect, mostly small cylindric, or oval; annulus differentiated and deciduous. Peristome lacking or rarely present and short; lid conic or rostrate, cells in straight rows; calyptra cucullate; spores small.

A genus represented in all the larger Continental areas and archipelagoes by seven-nine species, of which thirty-two, according to Brotherus, are known in America, twenty-two in South and five in Central America, four in the West Indies, and one in the United States.

Hyophila riparia (Aust.) Fleischer, M. S. in Austin Herb.

Pottia riparia Aust. Musci App. No. 112. 1870.

Trichostomum Warnstorfii Limpr. Laubmoose. **1** : 587. Fig. 171. 1888.

Leptodontium Canadense Kindb, in Mac. Cat. **6** : 45. 1892.

Leptodontium riparium Britt. in Bull. Torr. Club, **19** : 275. 1892.

Didymodon riparium Kind. Br. Eu. & N. Am. **2** : 280. 1897.

Plants ½ cm. high; stems branching, branches about 1 cm. long; leaves crowded when dry about ½ mm. long; cells square and small .005-.013 mm.,

Plate X. *Hyophila riparia*. This plate has been reduced one-half from the magnifications indicated for each figure.

1. Perichaetium dissected, showing one archegonium and four gemmiferous paraphyses, × 160.
2. Branched paraphyses with gemmæ, × 160.
3. & 4. Two paraphyses enlarged 340 diameters.
5. One of the gemmæ magnified 470 diameters.
   The other gemmæ show various forms and have been enlarged 340 diameters.

The July Bryologist was issued July 1st, 1904.

lower basal ones .012 x .027 mm.; apex serrate with irregular teeth; cells slightly papillose. Dioicous: antheridia in a terminal bud with small, blunt leaves. Archegonia few 1-10, paraphyses gemmiferous, branching into clusters of septate filaments bearing irregular multicellular gemmae, which produce protonema and serve as an asexual method of progagation! Seta 5-7 mm. long; capsule erect, 1 mm. long; lid conic; annulus large, double: spores not seen.

Rare in fruit, having been found but once by C. F. Austin "on rocks in a rivulet, Sussex County, N. J. September, 1867."

Illustration: Sullivant Icones Supplement, plate 21. 1874.

Habitat: "On moist rocks along streams" and shores of lakes.

Distribution: From Owen Sound, Ontario, to Pennsylvania and Ohio.

Type locality: "Palisades of Northern New Jersey and Southern New York." probably at Closter, N. J.

Also found by Austin at Hohokus, Pascack, and Little Falls, N. J.; Jordansville, Watkins Glen, Niagara Falls (Wolle), Chilson Lake (Mrs. Harris), Bashbish Falls (R. S. Williams), N. Y.; Springfield, Ohio (Miss Biddlecome), Bethlehem, (Rau & Wolle), Pocono Mt., Penn. (Porter), and at Owen Sound, Ontario (Macoun). Also in Europe on the shores of the Lake at Zürich, Switzerland, and at Schaffhausen on the Rhine.

The *gemmiferous paraphyses* which fill the perichaetial heads almost to the exclusion of the archegonia are characteristic of this species. The accompanying plate shows the diversity of form and size which occur in these gemmae. No mention of them has been made in the description of *Pottia riparia*, but in *Trichostomum Warnstorfii* they have been described and figured by Limpricht, Laubmoose 388. Figs. 170-171, and by Correns, Unt. Verm. Laub. 99. Figs. 38-40.

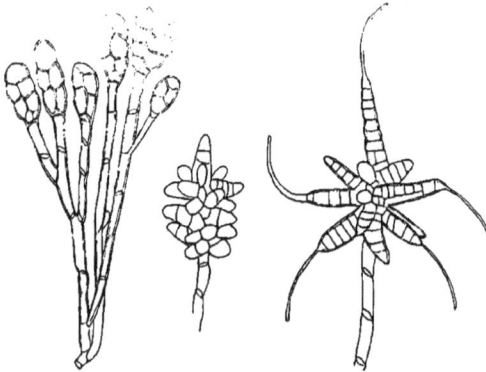

Trichostomum Warnstorfii.
Fig. 171. *Reduced from Limpricht.*

Through the kindness of Dr. Warnstorf I have received a specimen of *Trichostomum Warnstorfii* Limpr. (Fig. 171,) collected by F. Weber "on walls wet by spray along the shores of Lake Zürich, Switzerland." This is one of the cotypes figured by Limpricht in the Laubmoose. It has only been found sterile!

There appears to be considerable variation in the shape and serration of the leaves as well as in the development of the papillae in American specimens. In *Pottia riparia* (Aust. Musci App No. 112) the leaves are rounded, blunt at apex, like *Trichostomum Warnstorfii*, but slightly more serrate; *Leptodontium Canadense* Kind. has the leaves still more sharply serrate and the apex more acute; and in specimens collected by R. S. Williams, in the river below Bashbish Falls, the leaves are narrowly lanceolate, 2 mm. long and only ½ mm. wide and the serrations are few and very small, the vein also is serrate on the back at apex. These differences are of degree, however, not of kind, and seem hardly specific, as the branched paraphyses and multiform gemmæ are present in all, and although there are slight differences in the shape of the gemmæ, all the other characters are so much alike that unless some difference in the capsules were found it is more than probable that they are the same species.

Prof. Max Fleischer has described and figured a new species from Java as *Hyophila Dozy-Molkenboeri* Fl. (Die Musci d. Flora v. Buit. 1: 329 fig. 57. 1904) in which there are also polymorphous gemmae borne on the ends of the paraphyses. It grows on stones and earth on walls and roadsides in Java, Sumatra, and Japan? and Nepal?

New York Botanical Garden.

LICHENS—STEREOCAULON, PILOPHORUS AND THAMNOLIA.

Carolyn W. Harris.

Stereocaulon is a genus represented by a number of species, a few of which are quite common. The specimens are often very handsome, with their fruticulose, granulated podetia and numerous dark brown apothecia.

This genus belongs to the same family as the Cladonias, and resembles that genus in having a secondary thallus composed of stalk-like elevations called podetia. The primary thallus is deficient so that only a close observer will distinguish it. The secondary thallus is not hollow as in Cladonia, but is composed of arachnoid filaments of hypae, covered with a cartilaginous cortical layer on which are borne the gray branch-like granules known as phyllocladia. The podetia, especially in infertile specimens, are often covered with whitish soredia; these become squamose or pass into coral-like branchlets. The apothecia are usually numerous, in some species crowded, either terminal or lateral on the podetia; disk convex, becoming globose, dark reddish brown or nearly black.

The species of Stereocaulon are found largely in mountainous regions, either on sandy earth or on rocks. Their distribution is very general, being found throughout the world in the mountains; most of the species are found in North America.

When dry they are very brittle, but as they do not cling closely to the substratum can be collected at any time.

Stereocaulon coralloides Fr. The primary thallus, which is composed of coarse granules, soon disappears, and is usually not found in fully developed specimens. The podetia—or secondary thallus—are short, united at the base and are much branched toward the top. The branches are densely covered with coarse, light gray granules.

Apothecia are small, spherical, terminal or lateral, sometimes are solitary but oftener are crowded together at the end of the podetium; the disk is a dark, reddish brown, with no margin.

Found on rocks in the mountains of New England as well as those of the southern and western states.

Fig. 1. *Stereocaulon pas-chale* × 2.

STEREOCAULON PASCHALE (L.) Fr. Fig. 1. Primary t h a l l u s, as in *S. coralloides*, usually wanting. Podetia long, rather lax and slender, much branched and covered with squamulous, crenate, dark gray granules which are smaller than in *S. coralloides*. Along the podetia are frequently bare spots showing the broken cortical layer and the white medullary tissue inside.

Apothecia are at or near the apices of the podetia; they are rather small, very numerous with a flat, dark brown disk.

S. paschale is found on rocks and is a very handsome species. It grows in round, thick mats; the podetia are decumbent when dry, almost erect when moist. It is sometimes very abundant on rocks in cleared land on mountain slopes; is said to grow abundantly on volcanic lava.

STEREOCAULON TOMENTOSUM (Fr.) Th. Fr. Primary thallus distinct and granular. Podetia large, solitary or loosely caespitose, covered with crowded gray granules which are very tomentose. At the base the podetia are bare, especially on the under side.

Apothecia very small, terminal or lateral, disk dark brown, convex, becoming subglobose.

Found on the earth in mountainous regions,

S. tomentosum (Fr.) Th. Fr. var. *alpinum* Th. Fr. This is a small alpine form of *S. tomentosum*. The thallus is small, the podetia compressed and erect, not very tomentose, the granules are light gray, almost white and very powdery.

Apothecia usually terminal and rather large, with a dark reddish brown, convex disk.

Found on the ground or on boulders in alpine or sub-alpine districts.

Fig. 2. *Stereocaulon con-densatum* × 2.

STEREOCAULON CONDENSATUM Hoffm. Fig. 2. Primary thallus very small and warty, podetia very short, prostrate, with short branches which are covered with very crowded, powdery granules: these are from light to dark gray in color. The primary thallus often forms a granulose crust and the podetia almost disappear.

Apothecia terminal and in a close cluster, the disk convex, dark, reddish brown.

Found on the earth in upland districts, more frequently on or near the sea coast.

Pilophorus. This genus is represented by only one well marked species, although this has been divided into two or more varieties. It is a northern genus and not a common one. There is usually no apparent primary thallus, the podetia are rather rigid. cylindrical, slightly granular, especially at the base, are rarely branched. The color is a grayish or greenish yellow. The apothecia are terminal, round and black, solid within. The habitat is moist shady rocks in sub-alpine situations.

PILOPHORUS CEREOLUS (Ach.) Tuckerm. Fig. 3. Primary thallus usually wanting: if present is effuse, forming a granular crust: podetia simple, rarely branched, erect, pulverulent or granular, much wrinkled. Color greenish gray, sometimes almost yellow.

Apothecia rather large, nearly spherical, solitary, terminal, with a blue black disk.

Found on moist rocks in mountainous regions.

Thamnolia. This genus is generally placed in the same family with Cladonia and Pilophorus, although some lichenists do not agree with this classification. It is propagated by soredia, no apothecia having been observed. There is only one species which has been divided into two varieties.

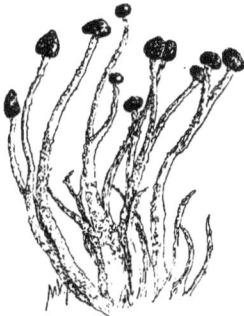

Fig. 3. *Pilophorus cereolus* × 2.

The primary thallus is wanting, the podetia—or secondary thallus—are long, very lax, slender and prostrate in some specimens, swollen and more erect in others, are wider at the base, tapering to a point at the top, they are somewhat wrinkled, are hollow throughout their length. The color varies from a pale straw color to a dull blue gray. Found on the ground among mosses in alpine and sub-alpine regions.

Thamnolia vermicularis (Sw.) Schaer. var. *subuliformis*. Secondary thallus simple or occasionally forked, usually wrinkled, prostrate and slender, long and very lax, a light straw color, sometimes turning to a grayish blue; is found on the ground in moist localities.

In var. *taurica* the thallus is more erect, less wrinkled and whiter than in var. *subuliformis;* it is swollen and quite pointed at the top.

T. vermicularis, Fig. 4, is easily recognized because of the peculiar form and the habitat of the thallus, which does somewhat resemble "white worms."

It is an alpine or sub-alpine plant and grows on the ground, frequently with mosses, looking much like some species of fungi.

Fig 4. *Thamnolia vermicularis* × 1.

CURBSTONE MOSSES.

Cora H. Clarke.

Having lately made an interesting discovery on a curbstone, I would suggest to those Chapter members who live in cities to examine the curbstones in the side streets, or those a little out of town where they are not too vigorously scraped and cleaned.

In the city of Meadville, Penn., where I have passed the winter, the moss-gathering capacity of the curbstones is sadly interferred with by the regrettable custom of raking up the autumn leaves and making bonfires of them, close to the sidewalks, whereby the mosses growing on the side of the curbstone are singed, blackened, or wholly killed, and sometimes the fire creeps through under a board walk, and destroys mosses growing on the stone base of an iron fence. I have such a piece of stone work in my mind, where I have found at different times a fruiting Amblystegium varium, two or three species of Bryum, Weisia viridula, Ceratodon, a Barbula, and Funaria hygrometrica.

In the part of the town where I stay, there is an edging of grass between the flagging and the curbstone, and the proximity of the grass seems favorable for the growth of mosses on both the top and the side of the curbstone. Funaria is the commonest species, and now, May 18, is very pretty, with its green cranes'-heads. Bryum argenteum is tolerably common, and some other Bryums, of which the heads are green still, with Ceratodon, and sundry pleurocarps which are not in fruit.

But the richest harvest of moss I found on the next street, on a curbstone facing north, and partly shaded by trees. The stone seems to be a sort of slate, but perhaps a little too smooth for the adherence of the mosses, for in many places they were lying in the gutter in a roll that had peeled off the curb. Where they were still attached, they looked like a solid green cushion, covering the whole side of the curbstone, and I was surprised to find them nearly all acrocarpus mosses. The silvery foliage of Bryum argenteum seemed quite concealed in the green growth, but the tiny bells hung here and there: the other Bryums sent up their needles—also Ceratodon; the Funarias showed both old brown capsules, and young shoots. It is my experience that in mosses, when the ripe fruit has shed its lids, the fruit of the next year begins to show the silvery sheen of its caps.

Patches of Barbula unguiculata grew here and there; most of the lids were gone at this date, April 18th, but the red peristome was in good condi-- tion. But my most interesting species was one that as I have hinted was new to me: here and there my eye was caught by a red gleam, which close inspection showed to be caused by tiny red capsules, each on a very short seta, and all the wee heads looking one way—a few still bore the lids, but most of them showed the red peristome. It was evidently a *Dicranella* and I have made it out to be *Dicranella varia*, a species which I never happened to find before, although I have some specimens in my herbarium.

In the hope that others may find species of equal interest to them on the curbstones to which they have access, I send this account of my experience.

Boston, Mass.

BOOK REVIEWS.

WARNSTORF, C. KRYPTOGAMENFLORA DER MARK BRANDENBURG UND ANGREN-ZENDER GEBIETE. ERSTER BAND. LEBER UND TORFMOOSE. 8vo. 481 pages, 231 figures in text. Gebrueder Borntraeger, Leipzig, 1903.

Herr Warnstorf, of Neuruppin, Germany, has long been known to both European and American botanists as an ardent student of the bryophytes and especially of peat-mosses. For many years he has devoted a great deal of attention to the rich bryological flora of the region around his home, and the present work is the result of this study. It will consist of two volumes. In the first, which is referred to above, the hepatics and peat-mosses are treated; the second will be devoted to the "Laubmoose" or true mosses and may be expected in the near future. The work is intended not only for the advanced bryologist but also for the beginner. To this end a chapter is included which gives directions for the collection, preservation and study of specimens, and there are also numerous and practicable keys for the determination of species. In the present volume, 111 species of hepatics and 39 of Sphagnum are accredited to the Brandenburg region. All of these species are described and all except a few are figured: attention is also called to many other species which are to be looked for in the region but which have not yet been found there. The descriptions are full and accurate and are often interspersed with biological, physiological and morphological notes of much importance and interest. The figures in most cases represent structural details, which bring out the differential characters of the species treated.

Herr Warnstorf's book is not merely of local interest, but has much in it of value to the American student, a fact which is at once apparent when we remember the close similarity between the bryological flora of Europe and that of northern North America. Among the Brandenburg hepatics, for example, no fewer than 85 have also been recorded from North America, while only 26 are peculiar to Europe. In New England at the present time, 128 species of hepatics are definitely known, 102 of which are common to Europe. Herr Warnstorf describes 70 of these species and alludes at length to 8 others: those which are not mentioned are mainly alpine or subalpine species of limited range. Among the Brandenburg Sphagna, all except 6 are common to North America. Aside from the purely descriptive portions of the work, the introductory chapter, which depicts the physiographical peculiarities of the region treated and the characteristic moss-societies to be found there, is especially to be recommended. Little work of this sort has as yet been done by American bryologists.

ALEXANDER W. EVANS,
New Haven, Conn.

BOOK REVIEWS—Continued.

THE MOSSES OF JAVA—" Die Musci der Flora von Buitenzorg " (Zugleich Laubmoosflora von Java) Max Fleischer. 2 vols. 8. vo. pp. + 643 XXXI, with 120 illustrations. E. J. Brill, Leiden, 1904.

These two interesting volumes contain not only many additions to the Flora of Java since the publication of The Bryologia Javanica in 1861, but they also emphasize the importance of the peristome in classification as may be seen by the Systematic Synopsis given in the preface, where the *Buxbaumioideae*, *Tetrapoidoideae* and *Encalyptaceae* constitutes separate orders and classes to which Prof. Fleischer has given new names. Thus far only the *Sphagnales* and *Bryales* have been issued, but these include many new species and original illustrations. The remaining volumes will be awaited with much interest.   ELIZABETH G. BRITTON,
New York Botanical Garden.

MUSCI IN ENGLER AND PRANTL'S NATURLICHEN PFLANZEN-FAMILIEN.

JOHN M. HOLZINGER.

This great work on the mosses of the world was begun in December, 1894, by Dr. Carl Müller Berolinensis, who, as he himself explains, must be distinguished from Carl Müller Hallensis. The Hepaticae are treated in pp. 3-141 by V. Schiffner. Musci begin on p. 142, and in the last number recently issued continue to p. 624, approaching the close of Acrocarpi, but Dr. Müller's contribution ceases on p. 202, comprising a very valuable enumeration of the most important literature, which covers twelve pages of closely printed lists of works, both early and recent, dealing both with the developmental history, the systematic arrangement, and the geographical distribution, this last arranged according to continents and countries.

This author's treatment from pp. 154-202 does not pass beyond the stage of the introductory discussion of the life history and of forms of propagation and the anatomical investigations of organs. On p. 203 Dr. Engler explains that the author by reason of pressure of official duties could not continue the elaboration of Musci for the Pflanzenfamilien, and that in consequence Dr. W. Ruhland will complete the elaboration of the propagation and life history of the Musci. Dr. Carl Warnstorf will contribute the systematic treatment of Sphagnaceae, and Dr. V. F. Brotherus that of the remaining families of Musci. Therefore from p. 203-243 Dr. Ruhland goes into great detail in discussing and illustrating by figures the propagation by sexual and vegetative methods of Musci and their life history. He also contributes an introductory chapter on general relations both to Sphagnales, to Andreaeceae and Bryales. Dr. Warnstorf then treats Sphagnaceae systematically, pp. 248-262. Dr. Brotherus covers Andreaeceae, on pp. 265-268, and Bryales from p. 277 on to p. 624, this last number of Pflanzenfamilien having recently appeared (April, 1904).

Like all the work in Engler & Prantl, the treatment of Musci is ably conceived, well carried out and beautifully illustrated with much microscopic detail, by the several authors. And especially is this true of the part that has been assigned to Dr. V. F. Brotherus; his is, in fact, the lions share of the systematic work and for this simple reason needs a somewhat more detailed review.

Brotherus, then, in the first six pages pp. 277-282, brings up to date (i. e. 1901) the enumeration of the most important moss literature of the countries of the world, supplementing that of C. Müller. In a delightfully concise introduction of less than half a page (p. 282) the author defines and defends his position regarding the systematic place of the Cleistocarpi, which C. Müller and Schimper separate as a natural group from the Stegocarpi, but which S. O. Lindberg considers as a lower stage of development of the latter. He holds Lindberg's view the more defensible here, as well as in the delimitation of families. In the treatment of both larger and smaller groups the author pays a deserved tribute to Prof. Limpricht, when he explicitly states that the latter's masterly treatment of European mosses has served him as an example in the treatment of anatomical characters in exotic forms. This in part explains the excellence of the many microscopical drawings in the work as far as completed. Nearly five pages, pp. 283-287, are covered by a skilfully arranged artificial Key to the Genera of Acrocarpous Mosses, including the Cleistocarpi and Stegocarpi, and this as well as the many keys to the species of the larger genera through the upward of 350 pages so far published, constitutes one of the most interesting and helpful and therefore most satisfactory features of this most able contribution to universal bryology,

In a subsequent note after the Acrocarpi are completed the writer will take pleasure in reviewing more in detail the author's treatment of this group.   Winona, Minn.

NOTES ON NEW OR RARE MOSSES

A Moss New to North America.

Dr. John W. Bailey recently collected a moss near Blackfoot, Idaho, at an altitude of 4,000 ft. above sea level, which from the description given in Limpricht's Laubmoose I, p. 524, appears to be *Pterygoneurum cavifolium incanum* (Bry. Germ.) Jur. Laubm. Fl. p. 96.

If my determination is right, this is an addition to our North American moss flora. The plant was collected in goodly quantity, and will be distributed in Fascicle 4 of my Acrocarpi.

JOHN M. HOLZINGER,
Winona, Minn.

Anacamptodon splachnoides, Brid.

Looking over Prof. J. Franklin Collins' " Notes on Mosses " Rhodora for August, 1903, I find that *Anacamptodon splachnoides* Brid. has not been reported from Connecticut. In the summer of 1899 I found it at Burnside, Conn., growing on a living Elm tree about five feet from the ground.

JOSEPHINE D. LOWE,
Noroton, Conn.

WHEN DOCTORS DISAGREE.

Dr. Grout has recently printed the following remarks on No. 196 of his North American Musci Pleurocarpi, which Cardot and Theriot have described in the May number of the Botanical Gazette as *Plagiothecium Groutii.*

"No. 196. This is a most interesting form about which "Doctors disagree" very widely. Mrs. Britton has called it a variety of *Plagiothecium denticulatum,* but I do not believe many others will agree with her view. M. Cardot has named it as above. Dr. Best is sure it is a depauperate form of *Raphidostegium recurvans.* "The perichaetial bracts, the pedical, the cilia, one or two, imperfect and shorter than the typical teeth, the imperfect annulus, the conical short-beaked operculum, about one-half the length of the urn and the quadrate-rectangular exothecial cells, are all about as we might expect in a starved form of *R. recurvans;* "only this and nothing more." Whatever else it may be it is not a *Plagiothecium!*" "I believe it is a new species which is more at home in *Raphidostegium* because of the enlarged alar cells in some of the larger plants. The smaller size, the shorter less slenderly acuminate leaves with alar cells much less strongly developed, complanate but not recurved and only very slightly unsymmetric, and the smaller capsules with shorter beak, seem to me so distinctive as to make this form worthy of specific rank. It hardly seems depauperate as it was growing with typical *R. recurvans* in a favorable spot (possibly a little dry at times) and besides it is fruiting freely. There is a little *R. recurvans* intertangled with it."

Now in this case there is every reason for disagreement because the specimens distributed as No. 196 were collected at Hempstead. Long Island, on December 1, 1899, and the specimens sent to me were collected at Lawrence, Long Island, May 25, 1899. I have recently examined *No. 196,* and am convinced that I have never seen these specimens before!

Furthermore they are mixed with *Rhynchostegium serrulatum* not *Raphidostegium recurvans,* and the cells of the basal angles are not inflated and only slightly differentiated, two or three being rectangular instead of long *prosenchymatous* like the rest of the leaf as figured by Theriot. There is no marked resemblance to any species of *Raphidostegium,* and it is incredible that Dr. Best should have mistaken it for a "depauperate form of *R. recurvans,*" especially as Dr. Grout admits that this species was "intertangled" with it! I have just compared these two species and *No. 196* has the flattened unequaled leaves of *Plagiothecium,* and the apex though serrate is much shorter and broader and not strongly recurved as in *R. recurvans!*

The description in the Botanical Gazette for May, p. 379, states that *P. Groutii* was collected in "Delaware: Hampstead"—this is also evidently a mistake according to Dr. Grout's label! The moral of this is that it is not worth while to print useless remarks about species that are mixed, nor to attribute wrong determinations of specimens which have never been sent!

ELIZABETH G. BRITTON,
New York Botanical Garden.

JAMES LAWRENCE BENNETT.

1832-1904

We take the following notice of the death of our friend, Mr. Bennett, by permission, from Rhodora 6: June, 1904: "James Lawrence Bennett, whose name has long been associated with the flora of Rhode Island, died at Hartford, Connecticut, April 30th, 1094. Mr. Bennett was born in Providence, April 8th, 1832. He was educated in the public schools of his native city and prepared for Brown University, which, however, he was unable to attend. For many years he was a manufacturing jeweller, but found time for scholarly pursuits and was widely read in the natural sciences. His keen interest in botany dated back at least to his twentieth year. His botanical collecting was done chiefly in Rhode Island, but extended to the White Mountains, which he visited about ten times. He made also smaller collections in northern Vermont, and in Tompkins County, New York. During 1890 and 1891 Mr. Bennett was curator of the Herbarium of Brown University, and from 1891 to 1894 the curator of the Herbarium and Museum of Economic Botany at the same institution. In 1891 he received an honorary degree of Bachelor of Arts from Brown University. In 1888 he published, under the auspices of the Franklin Society, his "Plants of Rhode Island, being an enumeration of plants growing without cultivation in the State of Rhode Island." This publication of 128 pages, dealing with both the flowering plants and several of the groups of cryptogams, is still the most comprehensive catalogue of Rhode Island plants. Mr. Bennett's herbarium of flowering plants has long been incorporated with the Herbarium of Brown University. It is said that his cryptogams were sold to the Brooklyn Institute.—Benjamin L. Robinson."

I belive the last statement to be incorrect. In the fall of 1901 I purchased a large part if indeed not all of Mr. Bennett's cryptogams, including several valuable Exsiccati. The Austin set Dr. Grout took, while I retained the Sullivant & Lesquereux; Wright's Cuban Mosses, a large collection of foreign mosses, sets from Hawaii, duplicates from many sources--several Exsiccati of Hepaticae, besides unnamed material. Mr. Bennett was evidently very fond of his collections, often referring to them in his letters as his little children, and he gave them into my keeping with the injunction to "love as well as care for them." About this same time I purchased the entire collection of Mr. D. A. Burnett, of Bradford, Keene Co., Pa., recently deceased, and presented a full set, something over five thousand species, to the Brooklyn Institute, and this may have caused the above misstatement.

ANNIE MORRILL SMITH.

AN ANSWER TO MRS. E. G. BRITTON'S LAST ARTICLE "NOTES ON NOMENCLATURE."

JULES CARDOT.

In the last number of the BRYOLOGIST 7: May, 1904, my amiable and learned colleague, Mrs. E. G. Britton, deals with me rather roughly for having substituted for *Brachelyma* Sch. the more ancient appellation of Cryphaeadelphus C. Müll. "This name, she says, besides being much less desirable than *Brachelyma*, is entirely misleading in its suggestion of relationship, and M. Cardot renders himself particularly liable to ridicule in view of the numerous sarcastic paragraphs published by him on nomenclature in his Revision of the types of Hedwig." That *Cryphaeadelphus* may be "less desirable" than *Brachelyma*, that it may be, besides "entirely misleading in its suggestion of relationship," does not prevent it from being twenty-five years older than *Brachelyma* and consequently to enjoy an absolutely unquestionable right to priority; and I am very much astonished at finding myself contradicted on the point by Mrs. Britton, so stubborn, as a rule, with regard to questions of priority.

I am very sorry for one thing; it is that I have rendered myself "particularly liable to ridicule." In spite of the shame brought upon me, I shall however ask Mrs Britton to observe that the passages of my Revision of the types of Hedwig and Schwaegrichen, which she maliciously hints at, hoping to make me contradict myself, refer to quite different cases. I have refused and always will refuse to revive, in order to substitute them for appellations that have long been in use, specific names applying to peculiar forms, not widely spread and not representing the type of the species, as is the case with *Barbula humilis* Hedw., and *Hypnum tenax* Hedw.; or names of doubtful species, "species incertae vel male conditae," concerning which it is impossible to agree, such as *Funaria Muehlenbergii* Hedw. fil. and *Orthotrichum coarctatum* Pal. Beauv.; or at last names that have fallen into complete oblivion, and have been superseded by other names usually and generally used for half a century, such as *Leskea adnata* Michx. 1803, known since 1851 to all writers by the names of *Hypnum microcarpum* C. Müll. and of *Raphidostegium microcarpum* Jaeg. Against such names, we can appeal to a fifty years prescription. But such is not the case with *Cryphaeadelphus*, which dates but from 1851 and which I substitute for a more recent name, that has been known for little more than a quarter of a century, and which has been, besides but little used, since even in 1884 Lesquereux and James united *Brachelyma* with *Dichelyma*.

In the same article, Mrs. Britton blames me for having forgotten to remark, in the March number of the BRYOLOGIST for 1904, that in the January number Louisiana was included in the range of *Papillaria nigrescens*. If Mrs. Britton is so kind as to take the trouble just to look at the end of my article, she will perceive it is dated November 18, 1903, and Dr. Grout will state to her that he received it some day of the foresaid month. Do what I would, it was then quite impossible I should point out the complementary note of my learned colleague, since that note was published only two months or so after I had written my article, and I had addressed it to the Editor of the BRYOLOGIST.

At last, Mrs. Britton observes I have wrongfully written *Pilotrichella cymbifolia* (Sulliv.) Ren. & Card., instead of *Pilotrichella cymbifolia* (Sulliv.) Jaegr. Most willingly I confess she is right. I should have looked into Jaeger's work, instead of merely and simply trusting to the article that has inspired mine and where you may read: *Pilotrichella cymbifolia* (Sulliv.) Ren. & Card. Musc. Amer. Sept. 44. 1895.(BRYOLOGIST 6:60). Now that article bears the signature of Mrs. Britton herself! Does not my amiable colleague fear to have rendered herself, in her turn, rather "liable to ridicule " whilst blaming me for a mistake she herself was the first to commit?

Last of all, I shall remark that Mrs. Britton gives wrongfully among the synonyms of *Homalothecium subcapillatum* Sulliv., *Pterogonium ascendens* Schw., and *Platygyrium brachycladon* Kindb. It is not know exactly what Bridel's *Pterigynandrum brachycladon* is. That author quotes as synonym of its species; *Pterogonium decumbens* Schw. Suppl. II. I. 32, Tab. CX, which from the description, the plate and the specimen preserved in Hedwig-Schwaegrichen's Herbarium, is obviously the *Homalothecium subcapillatum* (Hedw.) Sulliv. It is therefore possible Bridel's plant should likewise be related to that species. But certainly such is not the case with *Pterogonium ascendens* Schw. Supp. III. I. 2, Tab. CCXLIII, nor with the *Platygyrium brachycladon* Kindb. Eur. & N. A. Br. 31. Those two names concern one and the same species, with leaves provided with double and very short nerve, which has evidently nothing common with *Homalothecium subcapillatum* and which is, on the contrary, nearly related to *Platygyrium repens*, as I showed in my Revision of the Types of Hedwig and Schwaegrichen, with figures to support it. My opinion, based on the examination of the types of *Pterogonium ascendens* preserved in the collection of those two authors in Boissier Herbarium, has, besides, been admitted, without being discussed ever so little, by Mrs. Britton herself (BRYOLOGIST, 5:II) I do not know upon what reasons she now grounds her change of opinion; it seems to me it would be useful if she should state those reasons in the BRYOLOGIST.   Charleville, May 15, 1904.

HAMMOCK FORMATION.

The following is taken from the Journal of the New York Botanical Garden, Vol. V : August, 1904, p. 162.

The hammocks consist of isolated groups of hardwood trees, shrubs, vines and herbaceous plants in the pinelands. The dense, often almost impenetrable growth excludes the direct sunlight and maintains a high degree of moisture, both conditions being favorable to the development of fungi, hepatics, mosses and ferns, representatives of which occur in great abundance.   JOHN K. SMALL.

OFFERINGS.

[To Chapter Members only. For postage.]

Mrs. Agnes Chase, 59 Florida Ave., Washington, D. C. *Entodon seductrix* (Hedw.) C. Muell. c.fr.; *Atrichum undulatum* Beauv. c.fr. Collected in Illinois.

Mrs. J. D. Lowe, Noroton, Conn. *Plagiothecium turfaceum* Lindb., c.fr. Collected in Maine; *Dicranella rufescens* (Dicks.) Schimp., c.fr.; *Brachythecium plumosum* (Sw.) B. & S., c.fr.: *Tortella caespitosa* (Schwaegr.) Limpr., c.fr. Collected in Connecticut,

Mr. Charles C. Plitt, 1706 Hanover St., Baltimore, Md. *Mnium sylvaticum* Lindb., c.fr. Collected near Baltimore.

Mrs. M. L. Stevens, 39 Columbia St., Brookline, Mass. *Pogonatum alpinum* var. *arcticum* Brid., antheridial plants. Collected Mt. Holyoke, Mass. *Amblystegium fluviatile.*

Miss Annie Lorenz, 96 Garden St., Hartford, Conn. *Homalia Jamesii* Schimp. Collected Willoughly, Vt.

Mr. G. K. Merrill, Rockland, Maine. *Cladonia gracillis* Fr. var. *elongata* Fr.

Mrs. Carolyn W. Harris, 125 St. Mark's Avenue, Brooklyn, N. Y. *Stereocaulon paschale* (L.) Fr.

Miss C. M. Carr, R. F. D. Route 3, South Framingham, Mass. *Evernia furfuracea* (L.) Mann. Collected Sudbury, Mass.

BOTANICAL DRAUGHTSMEN

One of the illustrators of THE BRYOLOGIST is ready to do botanical drawing, microscopic or otherwise, by either the line or wash method.

Address Miss MARY V. THAYER,
Holbrook, Mass.

Biological drawing a specialty. Illustrations made for effective and truthful reproduction.

P. B. WHELPLEY,
Dublin, N. H.

—S3—

BOOK REVIEWS—Continued from Page 76.

HARRIMAN ALASKA EXPEDITION—CRYPTOGRAMS.

Of the scientific results of the Harriman expedition to Alaska in 1899 two instalments appeared some time ago. Three more have now been published almost simultaneously, by Doubleday, Page & Co. In Volume III, Grove K. Gilbert, of the United States Geological Survey, follows up Mr. Muir's earlier account of the glaciers with the story of his own observations and conclusions. A comparison is made between the size and conditions of rivers of ice as previously reported and as they existed in 1899, thus affording indications of the recent changes and furnishing the basis for future study. Professor Gilbert was especially alert, too, to phenomena which would throw light on the glaciation of the eastern part of the United States many thousand years ago.

The papers on "Geology and Paleontology" (Volume IV) were contributed by five different experts, three of whom accompanied the expedition, the others having discussed material submitted for examination. Dr. William H. Dall, Dr. B. K. Emerson and Dr. Charles Palache conducted their researches in person. Inasmuch as Mr. Harriman's steamer merely skirted the coast, geological investigation was necessarily fragmentary. So incomplete is existing knowledge about Alaska, though, that every addition counts. One of the noteworthy results achieved was the correlation of slates and shales in three widely separated regions, and the determination of their age as early Jurassic. Another was the discovery of molluscan fauna in Eocene rocks, in a locality that was the scene of volcanic activity in early Tertiary time.

The observations of living cryptogramic plants—fungi, lichens, algæ, mosses, sphagnums, liverworts and ferns—were made by Dr. William Trelease and several assistants, and are embodied in Volume V (Botany). No less than 1,616 species were found in all, and their relationship with others elsewhere has been carefully worked out. An entertaining account of the utilization of plants by the natives is contained in Dr. Trelease's introduction.

The fungi are treated by P. A. Saccardo and Charles H. Peck; the lichens by Miss Clara E. Cummins, with admirably simple keys; the algæ by Dr. Alton Saunders; mosses by J. Cardot and I. Theriot; sphagnums by C. Warnstorf, whose determinations have been edited by Dr. Trelease; the liverworts by Alex. W. Evans; and the pteridophytes by Dr. Trelease.

The phanerogams are to be presented in two volumes, under the editorship of Mr. F. V. Coville, and are announced for the current year.

A. M. S.

MOSSES

WITH HAND-LENS AND MICROSCOPE

QUARTO : bound in durable pamphlet cover. Printed on the best paper by the best Printers, the Mount Pleasant Printery.

The HAND-LENS CHARACTERS are fully emphasized so that these alone are often sufficient to determine the plant and in addition the MICRO-SCOPIC CHARACTERS .re treated in the same untechnical manner.

$1.00   part, post-paid.

Checks NOT on New York City MUST be made out for 10c. extra.

PART I.   86 pp.   CON1 .INS A COMPLETE TEXT-BOOK

ON MOSSES TREATING OF

Life-History, Structure, and Biology.   Illustrated Glossary, Revised and Rearranged Key to Families. Also a Systematic Treatment of the Families partly Through the Dicranaceae.

PART. II.   Ready May 15th.   DICRANACEAE TO ENCALYPTACEAE. 88 pp.   25 full page Plates.   50 cuts in the Text.

SEND FOR SAMPLE PAGES.

Published by the Author   A. J. GROUT, Ph.D., 360 Lenox Road, Brooklyn, N. Y.

Also a few sets of North American Musci Pleurocarpi.   Second century just completed. $7.50 per century.   A. J. GROUT.

VOLUME VII.   NUMBER 6

✳ NOVEMBER, 1904 ✳

THE BRYOLOGIST

AN ILLUSTRATED BIMONTHLY DEVOTED TO

NORTH AMERICAN MOSSES

HEPATICS AND LICHENS

EDITORS:

ABEL JOEL GROUT and ANNIE MORRILL SMITH

CONTENTS

Entered at the Post Office at Brooklyn, N. Y., April 2, 1900, as second class of mail matter, under Act of March 3, 1879.

Published by the Editors, 78 Orange St., Brooklyn, N. Y., U. S. A.

THE BRYOLOGIST

BIMONTHLY JOURNAL DEVOTED TO THE STUDY OF NORTH AMERICAN MOSSES
HEPATICS AND LICHENS.

ALSO OFFICIAL ORGAN
OF THE SULLIVANT MOSS CHAPTER OF THE AGASSIZ ASSOCIATION.

ubscription Price, $1.00 a year. 20c. a copy. Four issues 1898, 35c. Four issues 1899, 35c. ogether, eight issues, 50c. Four issues 1900, 50c. Four issues 1901, 50c. Four Vols. $1.50 Six issues 1902, $1.00. Six issues 1903, $1.00.

Short articles and notes on mosses solicited from all students of the mosses. Address manuscript to A. J. Grout, Boys' High School, Brooklyn, N. Y. Address all inquiries and subscriptions to Mrs. Annie Morrill Smith, 78 Orange Street, Brooklyn, N. Y. For advertising space address Mrs. Smith. Check, except N. Y. City, MUST contain 10 cents extra for Clearing House charges.

Copyrighted 1904, by Annie Morrill Smith.

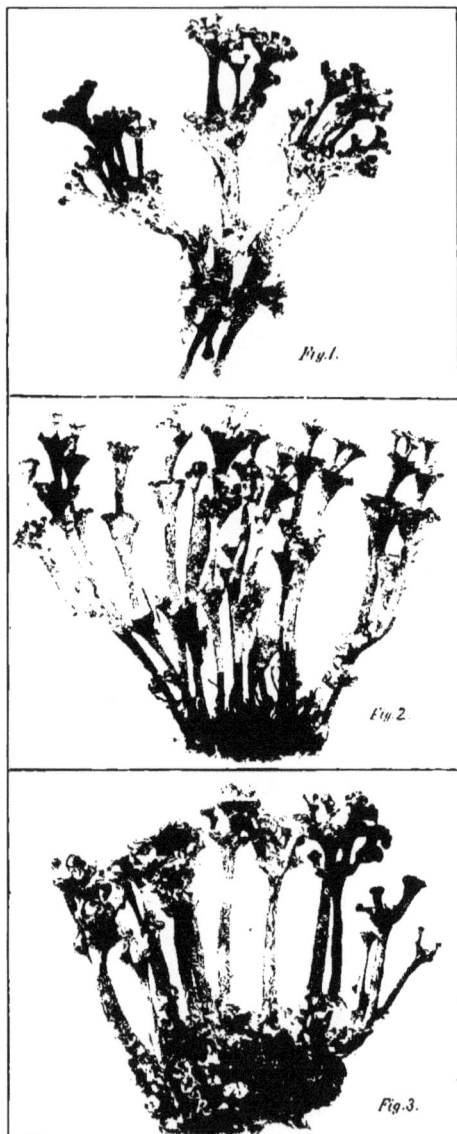

PLATE XI. Fig. 1. *Cladonia verticillata.* Fig. 2. *C. verticillata* var. *evoluta.* Fig. 3. *C. gracilis* var. *dilatata* × 1.

THE BRYOLOGIST.

Vol VII.   November, 1904.   No. 6.

FURTHER NOTES ON CLADONIAS.—IV.
Cladonia verticillata.
BRUCE FINK.

Tuckerman included *Cladonia verticillata* as a variety of *Cladonia gracilis*, but it is apparent enough that the plants that Tuckerman included in the variety are quite distinct from the others commonly placed under the species last named above. *Cladonia gracilis* is an exceedingly variable lichen, and it is doubtless true, as first appeared to the present writer after studying the forms found in northern Minnesota several years ago, that some European lichenists have carried the splitting process to extremes in their disposition of this species. The great degree of variation in the species is further shown in some very interesting forms that have come to me in the last few months, collected in New England by Mr. G. K. Merrill, and especially in that famous collecting ground in the White Mountains including "Tuckerman's Ravine." Also Mrs. Carolyn W. Harris has found some interesting forms in the Adirondacks.

But it seems best in disposing of *Cladonia verticillata* and *Cladonia gracilis* to dispose of the former species first as it is much easier to understand. and a presentation of figures and descriptions of the latter will form the basis of the next paper of this series. During the summer just passed, Dr. E. L. Harper, of Chicago, obtained some excellent photographs of lichens on Isle Royale in Lake Superior, and among the number were four or five of the two closely related species which are to receive attention in the present and the next following paper. I am under obligations to Dr. Harper for the photographs from which the illustrations in the present paper are taken. As readers of this series of papers are doubtless aware, it is not easy to bring out the characters of *Cladonias* in illustrations, but if the cuts can be made to bring out what is shown in the photographs, we shall succeed better than formerly. Then in the paper on *Cladonia gracilis*, we will be fortunate in being able to use photographs taken by Mr. Merrill from plants found in the region which has thus far proved richest in forms of *Cladonia gracilis*. In the present paper is given a most excellent likeness of *Cladonia gracilis dilatata*, Plate XI, Fig. 3, from a photograph by Dr. Harper, the object being to bring out the differences between the two closely related species.

Again we are able to have Dr. Wainio's view of a large number of specimens of the two species to be considered in this and the next paper. This is particularly fortunate regarding the various forms of *Cladonia gracilis*, the species to be disposed of in the present paper being far more easily understood as soon as one learns to distinguish it from the species named above.

The September BRYOLOGIST was issued September 2d, 1904.

Regarding the known distribution of *Cladonia verticillata*, we shall be able to make more definite statements than have been given concerning most of the *Cladonias* disposed of in previous papers. This is due to the fact that collectors and students have usually found the plant, and have not confused it with others so frequently as many other *Cladonias*. Indeed, one may depend on the lists of species published in the various states, for the distribution of this species, with a considerable degree of certainty. Yet we shall depend upon material actually seen, or the lists of a few lichenists of unquestionable ability. However, it is always unfortunate that a large amount of material in various herbaria can not be examined for the sake of additions to distribution. Finally, the forms of the species are not so numerous as in many *Cladonias*, and it is hoped that with the figures and the descriptions to follow, students of the *Cladonias* will have little further trouble with this species, except possibly the last two varieties, which are very rare. These we can not figure.

CLADONIA VERTICILLATA Hoffm. Deutschl. Fl. **2**:122. 1796. Plate XI. Fig. 1.

Primary thallus commonly persistent, composed of irregularly subcuneate, crenately lobed, or even incised-lobate, flat or somewhat involute, ascending, clustered or scattered, medium sized or larger squamules, which are 1.5–7.5 mm. long and wide, sea-green above or more commonly varying toward ashy, olivaceous or brownish, below white or darkening toward the base. Podetia arising from the lower margin of the squamules, 3–55 mm. long and .5–3.5 mm. in diameter, tubeaform or more rarely turbinate, sub-solitary or clustered into small patches, erect or rarely ascending, subcontinuous, grooved or areolate, with areoles usually closely contiguous, destitute of squamules, or rarely more or less squamose toward the base of the podetia or at the margins of the cups, sea-green varying toward ashy, yellowish, brownish, or olivaceous or these colors variegated, the narrow decorticate portions between the areoles white or rarely reddish, scyphiform. Cups medium sized or large, 2.5–9 mm. in diameter, usually abruptly dilated, shallow, the bottom closed or rarely cribrose, the margin subentire or dentate, commonly proliferous from the cavity of the cup, the proliferations one to several and the ranks usually two to five, the lowest rank about 20 mm· long. Apothecia small or medium-sized, .5–2.5 mm. in diameter, rounded or irregular, sometimes perforate, sessile on the margins of the cups, or short-pedicellate, flat and thinly margined or becoming convex and immarginate, paler or darker brown. Hypothecium pale or cloudy. Hymenium commonly pale below and brownish above. Paraphyses simple or rarely branched, commonly thickened and brownish toward the apex.

Found on various soils, both in shaded and open places, and frequently on thin soil over rocks or on decaying wood. Generally distributed throughout North America, except perhaps the extreme north and south; but more common or larger northward or in the mountains southward. Examined from several localities in New England and New York, and from Pennsylvania, Maryland, West Virginia, New Jersey, Ohio, Illinois, Iowa, Minnesota, Michigan, California, Missouri, Tennessee, Louisiana, Alabama, Florida, Newfoundland and several localities in British America. Macoun's " Catalogue of Canadian Plants " gives a wide distribution in British America. Also Clara E. Cummins has examined the plant from Alaska. Neither Tuckerman nor Wainio adds to this distribution. The specimens seen were collected by W. G. Farlow, Henry Willy, Clara E. Cummings, G. K. Merrill, E. A. Burt, Carolyn W. Harris, J. C. Eckfeldt, Emily Eby, T. A. Williams, H. A. Green, E. E. Bogue, Bruce Fink, E. L. Harper, H. E. Hasse, Colton Russell, W. W. Calkins, A. C. Waghorne and John Macoun. Known in all the grand divisions.

CLADONIA VERTICILLATA EVOLUTA Th. Fr. Lich. Scand. 83. 1871. Plate XI. Fig. 2.

Primary thallus commonly of smaller squamules. Podetia becoming elongated and consisting of several ranks, in ours commonly four to six. Examined by Dr. Wainio from my material from Minnesota, where the variety is distributed throughout the northern portion of the state. Habitat as above. Elsewhere examined by me from Isle Royale in Lake Superior; collected by Harper and figured herein; from Maine, collected by Merrill, from New Jersey, collected by Green; from the Adirondack mountains, collected by Mrs. Harris; and from Miquelon Islands, collected by Delamare. Nothing further can be definitely stated regarding the distribution of the variety, but it is probably common enough northward and in the mountains southward, and elsewhere in North America rare or absent. But we judge from Wainio's sequence of diagnoses and descriptions that he would give this form a general North American and foreign distribution.

CLADONIA VERTICILLATA CERVICORNIS (Ach.) Flk. Clad. Conn. 29. 1828.

Primary thallus persistent, composed of rather large or medium-sized, usually densely clustered, laciniate squamules, which are about 5-12 mm. long. Podetia rather short and slender for the species, 2-20 mm. long and .3-1 mm. in diameter, simple or proliferous from the central portions of the cups, or rarely from the margins or even from the sides of the podetia below the cups, the ranks 1-3, the upper ranks often without cups and branched irregularly, without squamules or squamose about the margins of the cups.

On humus among rocks or stones or in windy and sunny dry places, The only undoubted specimens seen are those collected in Germany by H. Sandstede and sent to me by the late Dr. F. Arnold, of München. However, another from our own country sent from Bay St. Louis, Mississippi, by A. B. Langlois and placed in the species by Hue, seems to be the variety. The whole plant is small, but the squamules are large in proportion, and the proliferations from the sides of the podetia frequently seen, and more rarely fruited ones from the margins of the cups. Wainio credits this form to Greenland, Arctic America, New Bedford and the White Mountains. Known in all the grand divisions. "Fere sicut *ovoluta* distributa est sed rarior," Wainio says. Wainio also states that his *Cladonia verticillata subcervicornis* Wainio

Mon. Clad. Univ. **2**:197. 1894, has occurred in Greenland. This is a smaller plant with shorter podetia, smaller squamules, one-or-two-ranked, and without squamules or proliferations on the podetia or the margins of the cups. We judge from Wainio's description that it is perhaps but an immature condition of var. *cervicornis*, and shall give it no separate description here.

CLADONIA VERTICILLATA ABBREVIATA Wainio Mon. Clad. Univ. **2**:197. 1894.

Primary thallus persistent, composed of smooth, laciniate, medium-sized squamules, which are 2-4 mm. long. Podetia arising from the upper surface or rarely from the margins of the squamules, about 1-1.5 mm. long and .3-.5 mm. in diameter, without cups and always terminated by apothecia, simple and without squamules, the cortex subcontinuous or rarely becoming areolate. Apothecia small, about 1-1.5 mm. in diameter, solitary or rarely aggregated at the summit of the podetium, flat and indistinctly margined by an exciple, or becoming convex and immarginate, brown or blackish brown.

On sandy earth. Wainio bases the variety upon material sent from New Bedford, Mass., by Henry Willey, and states that it passes into the normal form of the species. This variety is not known elsewhere, and I have not been able to examine it. However, if found elsewhere, there will be no difficulty in distinguishing it as it will probably occur in same environment with one of the better known forms of the species.

Cladonia gracilis, being the species with which the forms of *Cladonia verticillata* have sometimes been placed and with which they are easily confused, it has been thought best to give Plate XI, Fig. 3, Harper's excellent photograph of an average form from Isle Royale. The description will appear in the next paper of this series, and it is only necessary in closing this one, to give a few of the points of differences in the two species. In *Cladonia gracilis*, squamules are to be looked for anywhere on the podetia, while in *Cladonia verticillata*, they occur only at the base or on the margins of the cups. As a whole the podetial squamules are quite rare in the latter species, but common enough in the former. Also in the latter the proliferations are almost always from the central portions (or cavity) of the cup, while in the former they are nearly always from the margins. Without further differentiation, these may be regarded as the "ear marks" by which the two species may be distinguished. Other differences are less marked and are difficult to bring out even in the best descriptions.

Dr. Harper used excellent judgment in selecting and placing his specimens for photographing. It will be noted that they are all fruited, and it is to be hoped that the areoles and the decorticate lines between them will come out in the figures as well as they do in his likeness of var. *evoluta*.

Grinnell, Iowa.

HEPATICS WITH HAND-LENS.

A. J. GROUT.

There has been a considerable demand for a simple book on the Hepatics. To meet this demand I am preparing a treatment of the Hepatics similar to that which I have given the mosses in "Mosses with a Hand-Lens " This will be included in the second edition of that book now in preparation (See adv. in this No. of the BRYOLOGIST). This key to the genera is printed here with the hope that it will be used and criticised by the readers of the BRYOLOGIST, and that by the assistance of these criticisms the final treatment may be made more helpful. With the Queen ¼-inch achromatic triplet I am able to make out the more minute structures mentioned in the keys. Many of them, especially leaf structure, can not be made out satisfactorily unless the objects be mounted in water on a slide in the same manner as for a compound microscope. The slide should then be held up to the strong light, the slide being held with the left hand and the lens with the right, the right thumb resting upon that of the left hand so that the focus will not be distributed by any unsteadiness of the hands.

From now until winter closes in I shall be glad to attempt to name Hepatics for our subscribers if the speciments be accompanied by a stamp, full data for the label, and the best name the collector can give. Fresh material only is desired. Almost none of the books give the time of maturing spores of the different species, and I hope that our readers will send me all the data of this sort that they have. Comparatively few illustrations are possible in this article, but the figures in the sixth edition of Gray's Manual will prove very helpful. In working up this key I have been surprised to find that sterile Hepatics are, as a rule, much easier to identify than sterile mosses. Many of the species maturing their spores in early spring have the spores and capsules pretty fully developed in the preceding autumn so that some of the sporophyte characters are nearly always accessible. Hepatics shrivel more than mosses in drying and are best studied while fresh, especially the thalloid forms.

A few of the rare genera are omitted and in the completed treatment some of the minute or difficult species will not be included.

The Germans call the true mosses *Laubmoose*, meaning leafy mosses, and the Hepatics, *Lebermoose*, or liver mosses, The name Liverwort was originally applied to Marchantia because of its fancied resemblance to the liver. Because of this resemblance it was supposed to be a specific for all liver troubles according to the old doctrine of signatures. From this came the Latin name *Hepaticae* and the German *Lebermoose*. "Thus does the language of ignorant superstition become the adopted language of science."

The chief distinctions between Mosses and Hepatics have been noted in the BRYOLOGIST for April, 1899, but a few additional notes here may prove helpful.

The Hepatics may be leafy stemmed and appear much like mosses, or they may consist of a broad, flat and rather thin stem (thallus) which is usually closely applied to the substratum. These thalloid Hepatics might be mistaken for some of the foliaceous lichens but the Hepatics are always much greener and produce spores in a very different manner.

In the leafy-stemmed Hepatics, often called Scale Mosses, the leaves are without midrib and are nearly always in two ranks and flattened so as to lie in one plane, but in the great majority of cases there is a third rudimentary row on the inner side which are called underleaves, or amphigastra by those devoted to technical names. The pedicel which corresponds to the seta of the mosses does not, as a rule, grow much until the spores are nearly ripe, when it elongates very rapidly. The pedicels and capsules are of a much more delicate structure than in the mosses so that they disappear soon after the spores have escaped, but the peculiar and characteristic scales or bracts around the base of the pedical often remain much longer and help greatly in identifying species. Immediately surrounding the base of the pedicel is a tubular, somewhat three-sided organ called the inner involucre or perianth, surrounding this the outer involucre, called simply involucre by many authors. This latter may be either tubular or composed of separate leaf-like divisions of varied shapes, called involucral leaves or bracts, or perichatial leaves or bracts, or simply bracts. Either one, or even both, of these involucres may be lacking in some species.

So far as possible gametophyte characters have been used in the keys and descriptions and in the great majority of cases identification is easy from this part of the plant alone. Hepatics generally grow in moist situations on soil, roots of trees, and decaying wood.

Key to Families.

Plants leafy, mosslike in appearence except for the two-ranked leaves entirely lacking midrib......... .....Scale Mosses (Jungermanniaceæ).
Plants consisting of a flattened green thallus, sometimes nearly circular but usually elongated and branching. (See illustrations of Riccia, Marchantia, Anthoceros, etc.) ....A.

A.

1. Capsules, if present, immersed in the tissue of the plant. Plants floating on the surface of still water or floating on the mud along the banks...Riccia.
Capsules raised well above the thallus. Plants often growing in mud but never floating....................... ....2.

2. Stomata (in our genera) present, easily discernable with a lens as small pores on the upper surface of the rather thick thallus; capsule borne on a special stalked receptacle as in Marchantia.
Liverworts (Marchantiaceæ).
Stomata not present, on the thinner thallus; capsules never borne on a special stalked receptacle...3.

3. Capsules very long and slender, splitting into two valves when ripe after the manner of a mustard pod, the slender hairlike columella remaining in the center..... ....Horned Liverworts (Anthocerotaceæ).
Capsules globular or ovoid, splitting into four valves; columella lacking.
Thalloid Scale Mosses (Metzgeriaceæ).

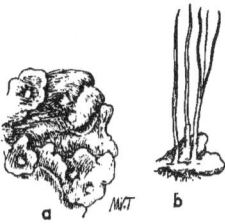

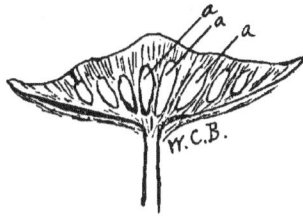

FIG. 1.　　　　FIG. 2.　　　　FIG. 3.

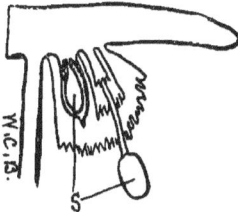

FIG. 4.　　　　FIG. 5.　　　　FIG. 6.

PLATE XII.

Fig. 1. a, Sterile and b, fertile thallus *Anthoceros punctatus* × 2 & 1.
Fig. 2-6. *Marchantia polymorpha*, from BRYOLOGIST, 4:34-35, 1901. Fig. 2. Male plant a little reduced, showing antheridial receptacles. Fig. 3. Longitudinal section of antheridial receptacle magnified. Fig. 4. Female plant reduced showing the stalked receptacles which characterize this family. These receptacles vary in the family from the shape shown in this figure to almost perfectly conical and entire. Fig. 5. Section of a part of a female receptacle magnified, showing two sporogonia. The seta of one has elongated, pushing the capsule out from the outer fringe (involucre) and the inner fringe (perianth) at the base of the seta is a little collar representing the base of the broken calyptra. Fig. 6. Sterile thallus with gemmæ.

THE TRUE LIVERWORTS (MARCHANTIACEÆ).

The plants of this family consist of thallus of medium to large size, one-half to six inches in length. usually branching dicotomously but sometimes with more than two branches at a fork. They are attached to the substratum by numerous roothairs and are thickened in the middle to form a midrib. This in some cases is not vary apparent above but shows plainly underneath. The upper surface is covered with small pores (stomata) which are very apparent with a lens, except in Reboulia. The capsules are spherical or ovoid and open irregularly by imperfect valves or by a portion of the top coming off after the manner of a lid. In this family the capsules and usually the antheridia are borne on special long-stalked receptacles well illustrated by the familiar Marchantia.

Key to the Genera.

1. Sterile stems bearing abundant gemmæ in shallow open receptacles....2
Sterile stems without gemmæ 3
2 Found only in and around greenhouses: gemmæ in crescent-shaped receptacles; never fruiting in our region...................Lunularia.
Growing abundantly everywhere; gemmæ in cup-shaped receptacles: capsule-bearing receptacle with 7-11 conspicuous rays..........Marchantia.
3. Thallus large; 2-6 inches long and ½ inch or more wide, distinctly areolate as in Marchantia, but areolæ larger and hexagonal... Conocephalus.
Thallus less than two inches in length and much narrower.............. 4
4. Pores (stomata) scarcely distinguishable; antheridia in sessile receptacles which might be mistaken for gemmæ; thallus purple on the margins; midrib strong underneath but not conspicuous above.........Reboulia.
Pores conspicuous, white: antheridia in peduncled disk-like receptacle; thallus with numerous dark purple scales underneath.... ....Preissia.
Pores conspicuous; antheridia immersed in the thallus; thallus purple underneath, at least along the margins 5
5. Perianth conspicuous, split into 8-16 fringe-like lobes; peduncle not chaffy............................... ....Asterella.
Perianth lacking: peduncle chaffy at top and bottom.............Grimaldia.

(The Reboulia of this key is the Asterella of Gray's Manual and the Asterella is the Fimbriaria of that work.)

THE THALLOID SCALE MOSSES (METZGERIACEÆ).

The spore bearing portion of plants of this family is like that of the Scale Mosses, but the green part of the plant is a thallus instead of a leafy stem in nearly all cases. There are, however, some intermediate forms in the family in which the thallus is divided into leaflike lobes. The thallus is much less highly differentiated than in the Liverworts and Riccias: there are no areolæ or pores (stomata), and the thallus is much thinner than in the Liverworts, in some species consisting of only a single layer of cells except at the midrib. The capsules are borne singly on setæ arising directly from the thallus. They are spherical to elongated-ovoid and remain enclosed in the calyptra until mature when the setæ rapidly elongate and break open the calyptra which is left at the base of the seta. The capsules open by four valves as in many of the Scale Mosses. A careful search of wet bare earth in shaded or springy places will nearly always yield one or more species of this family.

Key to the Genera.

1. Thallus with a distinct midrib ....2.
Thallus without a distinct midrib...........,.......................4.
2. Thallus 1/25 to 1/12 inch wide, dichotomously branched, cilliate along the margins...Metzgeria.

Thallus ⅓ to ½ inch wide, not cilliate at margins, entire or lobed.........3.

3. Thallus simple or only once forked, 1 to 4 inches long, prostrate; margins
    sinuate to entire; capsule ovoid-cylindric...................Pallavicinia.

Thallus dichotomously branched, ¾ to 1½ inches long, often densely clustered and ascending, margins lobed; capsules spherical without perianth,
appearing buried in the midrib for some time before the ripening of spores (Fig. 7)..................................Blasia.

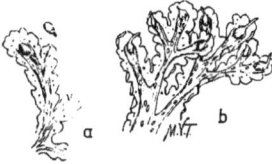

4. Thallus pinnately or palmately branch-
    ed, 1/24 to 1/12 in wide (except
    R. pinguis).................... Riccardia.

Thallus subsimple or dichotomously
branched, ⅛ to ⅓ inch in width (Fig.
8)................................. Pellia.

Fig. 7.

Fig. 7. *Blasia pusilla* L. a. Fertile plant in August
showing capsule in position. At the side is shown the capsule removed from the thallus. b. Sterile plant with flask-
shaped bodies which produce gemmæ.

Fig. 8. *Pellia epiphylla* Raddi. Thallus × 1, showing
involucre and position of capsule as it appears in August.

Fig. 8.

THE SCALE MOSSES (JUNGERMANNIACEÆ).

The reproductive part of the Scale Mosses, including the ripened capsule and its connected parts, perianth, involucre, etc., is essentially as in
the Thalloid Scale Mosses, but the vegetative part strongly resembles the
true mosses in general appearance. The leaves, however, are apparently
flattened out into two rows, one on either side of the stem. They are
entirely without midrib and are frequently two-cleft or lobed. One of the
lobes is often smaller and folded under the other making the leaves " complicate-bilobed," in the language of the books as shown in the illustrations
of Radula and Porella. This can best be made out by holding a single
stem up to the light and examining with a lens, when the under lobe will
show plainly as a deep shadow. In Scapania, the under lobe is the larger
and the plants look as if there were four rows of leaves. The lower lobe is
called the lobule and the upper simply the lobe. Very many species have a
third rows of leaves on the under side of the stem called technically " amphigastra" or underleaves, these vary in size from one-third the size of the
ordinary leaves to so minute that high powers of the compound microscope
are needed to see them clearly. The upper margin of the leaves may overlap the lower margins of the leaves next above as in Porella, or the upper
margin of a leaf may lie under the lower margin of the leaf next above as
in Plagiochila. In the former case the leaves are said to be incubous, in
the latter succubous. As this distinction is in most cases easy to observe,
it is given a prominent part in the key. Occasionally the leaves are so far
apart that it is hard to determine the leaf arrangement, but a careful search

will usually discover some plants in which this character can be seen. In plants with incubous leaves the bud is turned downward; when the leaves are succubous the terminal bud is turned up. So far as possible the key has been based upon the leafy or vegetative portions of the plants, but in some few cases the characters connected with the reproductive organs and capsules are necessary to accurately determine a plant. In most cases the characters used can be determined without mounting, if, however, they can not be readily made out the parts should be mounted as for the compound microscope. If one has access to a compound microscope it will often prove a very great help, although not necessary to make out the characters mentioned. Mnium and Fissidens are sure to be mistaken for Hepatics by the beginner unless the midrib or the leaves is noted.

Key to the Genera and Species.

1. Leaves entirely or in large part composed of hair-like divisions (easily observed if held up towards a strong light)....................2.
Leaves not as above ....3.
2. Plants grayish green, growing over the ground amid mosses in cool bogs, at least twice pinnate and somewhat resembling the Fern Mosses; leaves divided to base into hair-like lobes ....Trichocolea.
Plants dark green, much smaller, growing chiefly on rotton wood, but also found on humus-covered stones and soil; leaves with a considerable solid portion...Ptilidium.
Plants exceedingly minute, looking like a small green alga or moss protonema. Scarcely recognizable except when fruited; common on decayed wood, moist soil, etc............................. Blepharostoma.
3. Leaves incubous...A.
(Scapania and Chiloscyphus forms may be sought here.)
Leaves succubous...B.

A.

1. Leaves complicate-bilobed, upper lobes entire or nearly so (except Jubula). See figures and description of Porella...................2.
Leaves sometimes lobed or cleft but not complicate-bilobed................5.
2. Plants blackish or brownish green, minute, leafy stems 1/25 inch or less wide; lobule like an inflated sac (Plate XIV) ....Frullania.
Plants often dark olive-green but not often blackish; 1/16 inch in width, lobule not sac-like ....3.
3. Under leaves lacking; perianth strongly flattened crosswise (Plate XIII)...Radula.
Underleaves conspicuous .... 4.
4. Lobule with its longer edge attached to lower margin of lobe (See cuts, Plate XIII)..Lejeunea.
Lobule with its shorter margin attached to the lower edge of lobe (Plate XIV)...Porella.
5. Leaves mostly entire ....Kantia.
Leaves strongly toothed, notched, or cleft at apex...................... 6.
6. Leafy stems less than 1/25 inch in width.....................Lepidozia.

Leafy stems 1/16 to ¼ inch in width, with downward growing stolons
(Plate XIII)...Bazzania.

B.

1. Leaves complicate-bilobed, lobes nearly equal or the lower larger giving the appearance of four rows of leaves of which the two upper are
incubous and the two lower succubous......................Scapania.
Leaves not complicate-bilobed, in some cases toothed or divided.........2.
2. Leaves undulate on the margin; plants densely clustered; roothairs
bright claret colored...............................Nardia hyalina.
Leaves entire or slighly emarginate; roothairs colorless.................3.
Some or usually all of the leaves strongly toothed or lobed; roothairs colorless..................... ....7.
3. Leafy stems at least ⅛ inch wide, leaves plainly overlapping, on ground
and over mosses...5.
Plants about ⅛ inch wide: many leaves not overlapping.................4.
Leafy stems 1/16 inch wide or less..............................6.
4. Aquatic, floating, underleaves absent.
Chiloscyphus polyanthus var. rivularis.
On old logs and moist ground, underleaves present..........Chiloscyphus.
5. Plants creeping; leaves oblong to oblong-ovate, decurrent....Liochlæna.
Plants ascending; leaves round obovate, not decurrent........Plagiochila.
6. Leaves with a border of larger cells which appear as a whitish margin
under the lens..................Odontoschisma and Nardia crenulata.
Leaves without border of larger cells.............Jungermannia Schraderi.
7. Upper leaves with a strongly many toothed margin........Plagiochila.
Leaves 3-5 cleft ....Jungermannia barbata.
Leaves two toothed or cleft..... ....8.
8. Plants minute, leafy stems less than 1/25 inch wide; underleaves absent
or so small as to be invisible with a lens; leaves round-ovate to
obovate, cleft for at least ¼ their length.................. Cephalozia.
(Some small species of Jungermannia may be sought here but their
leaves are less deeply cleft and the plants are a much darker green).
Leafy stems at least 1/16 inch wide; leaves two toothed but scarcely cleft .9
9. Underleaves ⅓ the size of the other leaves..........Lophocolea minor.
Underleaves absent or minute............................... 10.
10. Leaves subvertical, varying from bidentate to retuse, or even entire
near apex................................... Lophocolea heterophylla.
Leaves with the edge attached nearly lengthwise of the stem. extending out at
almost right angles from it and lying flat in a horizontal plane. Geocalyx.
Leaves subvertical and all alike inserted more nearly crosswise of the
stem...11.
11. On sterile ground in open woods.............Jungermannia excisa.
On rotton wood...12.
12. Leaves round-ovate with an obtuse sinus between the teeth. Harpanthus.
Leaves subrectangular with an acute sinus; plants dark to brownish
green...........................Jungermannia Michauxii.

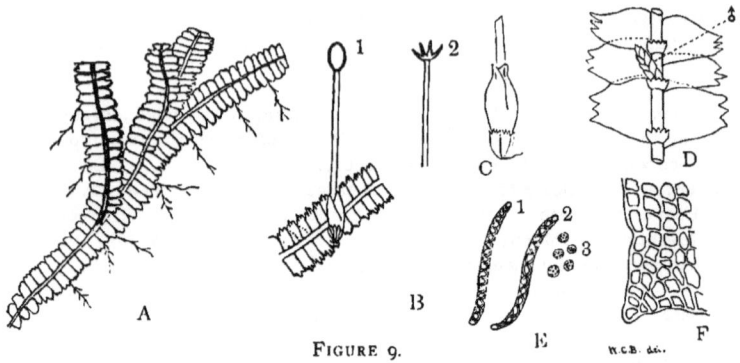

FIGURE 9.

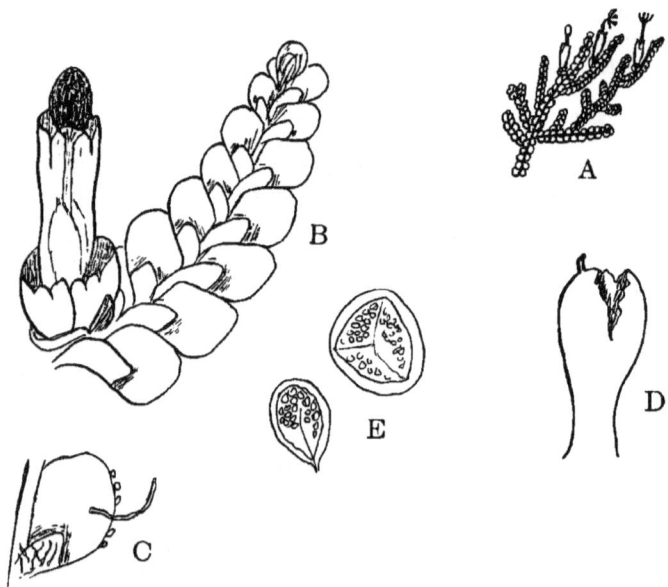

FIGURE 10.

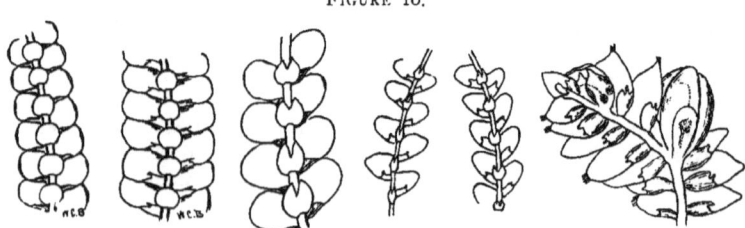

FIGURE 11.

PLATE XIII. *Bazzania, Radula, Lejeunea.*

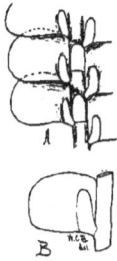

FIG. 12.

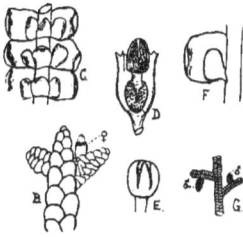

FIG. 13.

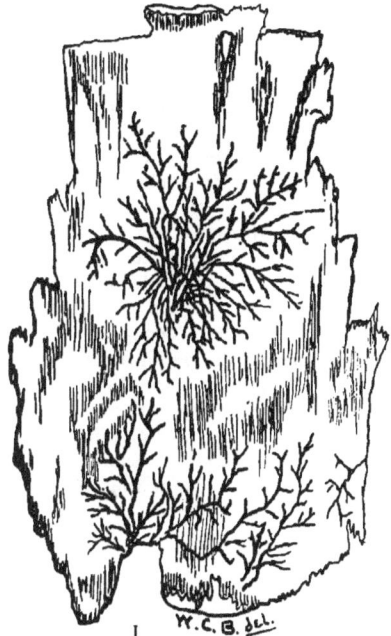

I.

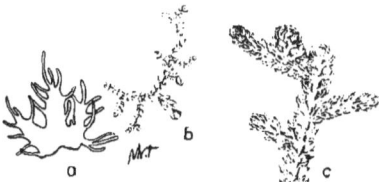

FIG. 15.

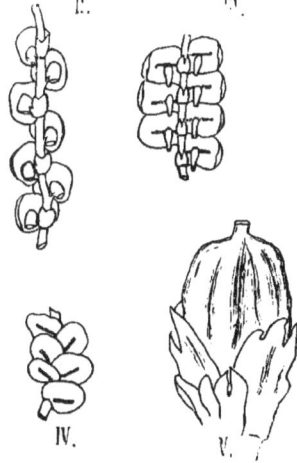

IV.

FIG. 14.

PLATE XIV. *Porella, Frullania.*

EXPLANATION OF PLATES.

Plate XIII. Fig. 9. *Bazzania trilobata* (L.) S. F. Gray. From BRYOLOGIST, 4: 68, 1901. A. Plant slightly magnified showing flagella springing from the underside. B. (1.) Portion of female plant with capsule. (2.) Capsule open. C. Involucre, perianth and base of seta enlarged. The involucre consists of the small leaves at the bottom of the figure. D. Male plant seen from below, showing antheridial branch, minute underleaves and incubous arrangement of leaves. E. & F. Illustrate spiral elators, spores and cell structure of leaf which cannot be seen clearly with a hand-lens.

Fig. 10. *Radula complanata* Dumort. A. Plant natural size. B. Branch with fruit showing clearly the seta and capsule, with the calyptra at base of seta showing through the transparent tubular perianth, and at base of the perianth, the involucre. This misrepresents the leaves, making them appear succubous. C. Leaf showing lobule with roothairs and larger lobe with gemmæ along the edge. This illustrates the simplest forms of "complicate-bilobed" leaf. There are no underleaves. D. Calyptra. E. Spores, highly magnified.

Fig. 11. Various species of *Lejeunea*, from BRYOLOGIST, 6:-27, 1903. Showing underleaves in all but the right hand figure. Note that the lobule is attached to the lobe by its longer edge.

Plate XIV. Fig. 12. *Porella pinnata* L. From the BRYOLOGIST, 5: 34, 1902. A. Underside of stem showing narrow underleaves and narrow lobules attached by their shorter edge to lobe. B. Single leaf showing lobe and lobule.

Fig. 13. *Porella platyphylla* (L.) Lindb. From BRYOLOGIST, 5: 35, 1902. B. Upper side of stem showing perianth and emerging capsule. Also showing clearly incubous arrangement of leaves. C. Underside of stem, the leaves shown too far apart. D. Longitudinal section of perianth. E. Capsule. F. Leaf. G. Part of plant showing male branches.

Fig. 14. *Frullania*. From BRYOLOGIST, 5: 4, 1902. I. Plant of *Frullania Eboracensis* Gottsche., on the bark of birch. II. Underside of same showing underleaves and the queer saclike inflated lobules which remind one of the bladders of Utricularia. III. and IV. Under and upper side of *F, Asagrayana* Mont. V. Involucre and perianth of *F. Eboracensis*.

Fig. 15. *Ptilidium ciliare* (L.) Nees. a. Leaf × 37. b. Plant with perianth and young capsule × 2. c. Portion of plant × 5.

SULLIVANT MOSS CHAPTER NOTES.

Students of the Hepaticæ will be pleased to know that Miss Caroline C. Haynes, 16 East 36th Street, New York City, will in the future have charge of the Chapter work and Herbarium and all members are requested to communicate with her at the above address.

IMPORTANT NOTICE.

There will be a meeting of The Sullivant Moss Chapter during the last week of December, 1904, in Philadelphia, in connection with the meetings of the American Association for the Advancement of Science. Information as to date and place of meeting will be sent to each member by postal card as soon as arrangements are perfected. Members are urged to show their interest in the Chapter by a large attendance. A number of papers will be read, and there will be exhibit of Mosses, Hepatics and Lichens from the several Chapter Herbaria and also from local collections. All are invited to contribute rare or beautiful specimens of Mosses, Hepatics and Lichens, or anything else illustrative of work accomplished by Chapter members, which will add to the interest of the meeting. Place of meeting will be the Buildings of the University of Pennsylvania, date between the 28th and 31st of December, 1904.

For further information address the secretary,

MISS MARY F. MILLER,
1109 M. street, N. W., Washington, D. C.

THE STUDENT'S HANDBOOK OF BRITISH MOSSES. BY DIXON & JAMESON. SECOND EDITION.

During the summer we received a copy of the second edition of Mr. Dixon's valuable work. There are five new plates and descriptions of thirty additional species and subspecies, besides numerous varieties. There are also many additions to the notes of Edition I. As we have before stated the descriptions in this work are models of their kind. The greatest need of America bryology at the present time is a similar work including and illustrating all our North American species.   A. J. G.

MUSCI ACROCARPI BOREALI-AMERCANI—Fascicle IV.

Nos. 76-100 of Prof. Holzinger's exsiccati have just reached us. Besides the regular numbers several supplementary numbers are issued with this fascicle.

Among the interesting things are Archidium Ravenelii Aust., Ceratodon minor Aust., Rhabdoweissia fugax (Hedw.) B. S., Trematodon longicollis Rich., Pterygoneurum cavifolium incanum (Bry. Germ.) Jur., Trichostomum tophaceum Brid., Orthotrichum fallax Sch., O. Hallii Sull. & Lesq., Physcomitrium Hookeri Hpe., Philonotis radicalis (P. B.) Brid., Webera Lachenaudii Cardot, Webera Tozeri (Grev.) Sch.   A. J. G.

NOTICE—ELECTIONS FOR 1905.

Members are requested to forward their ballots AT ONCE to the Judge of Elections, Mrs. Agnes Chase, 59 Florida avenue, N. W., Washington, D. C. Polls open until November 30th. The following candidates have been nominated, but if the members prefer to vote for a different set of officers, they are at perfect liberty to substitute any name or names on the list of members, in place of those given below: The members receiving the highest number of votes will be declared elected.

For President—Mr. Edward B. Chamberlain, 1830 Jefferson Place, Washington, D. C.

For Vice-President—Mrs. Carolyn W. Harris, Hotel St. George, Brooklyn, N. Y.

For Secretary—Miss Mary F. Miller, 1109 M. street. N. W., Washington, D. C.

For Treasurer—Mrs. Annie Morrill Smith, 78 Orange street, Brooklyn, N. Y.

OFFERINGS.

[To Chapter Members only. For postage.]

Mr. Walter Gerritson, 66 Robbins street, Waltham, Mass. *Sphagnum cymbifolium* (Ehrh.) Hedw.. c.fr.: *Leptobryum pyriforme* (L.) Wils., c.fr. Collected in Waltham, Mass.

Mrs. Mary L. Stevens, 39 Columbia street, Brookline, Mass. *Heterocladium squarrosulum* (Voit) Lindb., c.fr. Collected in Laconia, N. H.

Mrs. J. D. Lowe, Noroton, Conn. *Thuidium paludosum* Sulliv.) Rau & Hervey. c.fr.: *Plagiothecium striatellum* Lindb., c.fr.; *Thuidium Virginianum* Lindb. c.fr.; *Hypnum reptile* Mx., c.fr.; *Amblystegium irriguum* (Hook. & Wils.) B. & S., c.fr. Collected in Connecticut. (Mrs. Lowe has offered the first three of these mosses before and now offers them again for the benefit of new members, though others are welcome to a fresh supply.)

Miss Alice L. Crockett, Camden, Maine. *Cladonia cristatella* Tuckerm.; *Endocarpon fluviatile* D. C. Collected in Camden, Maine.

Mrs. R. E. Metcalf, Hinsdale, N. H. *Cladonia turgida* (Ehrh.) Hoffm. Collected in Hinsdale, N. H.

Mr. G. K. Merrill, 564 Main street, Rockland, Maine. *Thamnolia vermicularis* (Sw.) Schaer. Collected Mt. Washington, N. H.

Mrs. Annie Morrill Smith, 78 Orange street, Brooklyn, N. Y. *Antitrichia Californica* Sulliv. & Lesq., c.fr.; *Hylocomium loreum* (L.) B. S., c.fr. *Hylocomium triquetrum* (L.) B. & S., c.fr. Collected by Mr. A. J. Hill British Columbia.

THE BRYOLOGIST

AN ILLUSTRATED BIMONTHLY

DEVOTED TO

NORTH AMERICAN MOSSES

HEPATICS AND LICHENS

VOLUME VIII. 1905

EDITORS

ABEL JOEL GROUT AND ANNIE MORRILL SMITH

193922

PUBLISHED BY THE EDITORS

78 ORANGE STREET, BROOKLYN, N. Y.

INDEX 1905

SUBJECT INDEX

AUTHOR'S INDEX

ERRATA

Page 1, line 13 from bottom, for *Sullivantae* read *Sullivantiae*.

Page 2, line 18 from bottom, for Europeae read Europaea.

Page 3, line 10 from bottom, for Europea read Europaea.

Page 6, line 15, for ONITHOPODIOIDES read ORNITHOPODIOIDES.

Page 7, line 21, for INTEGROFOLIA read INTEGRIFOLIA.

Page 25, line 12, for iodine read iodide

Page 43, line 7 from bottom, for Sphaerocaphalus read Sphaerocephalus.

Page 51, line 22, for saxitalis read saxatilis.

Page 53, line 15, for *endivaefolia* read *endiviaefolia*.

Page 53, line 18, for *Lyelii* read *Lyellii*.

Page 54, line 11, for *leavis* read *lævis*.

Page 57, line 3 of Explanation of Plate V, for *asplenoides* read *asplenioides*.

Page 71, line 2, for **M. C.** read **C. M.**

Page 80, line 10, for **n. sp.** read **nom. nov.**

Page 94, line 8 from bottom, for *cylindrothecium* read *cladorrhizans*.

Page 102, line 8 of key, insert b before Ap. scarlet or orange.

Page 102, last line, for *Everina* read *Evernia*.

Page 103, line 3, for *Everina* read *Evernia*.

Page 103, line 15, for fibrilose read fibrillose.

Page 104, line 18, for filbrillose read fibrillose.

Page 106, line 5, for verraculose read verruculose.

Page 109, line 16 from bottom, for thallus read talus.

Page 112, line 12, for Floerk's read Floerke's.

VOLUME VIII.   NUMBER 1

JANUARY, 1905

THE BRYOLOGIST

AN ILLUSTRATED BIMONTHLY DEVOTED TO

NORTH AMERICAN MOSSES

HEPATICS AND LICHENS

EDITORS:

ABEL JOEL GROUT and ANNIE MORRILL SMITH

CONTENTS

Entered at the Post Office at Brooklyn, N. Y., April 2, 1900, as second class of mail
matter, under Act of March 3, 1879.

Published by the Editors, 78 Orange St., Brooklyn, N. Y., U. S. A.

PRESS OF MC BRIDE & STERN, 97-99 CLIFF STREET, NEW YORK

THE BRYOLOGIST

BIMONTHLY JOURNAL DEVOTED TO THE STUDY OF NORTH AMERICAN MOSSES
HEPATICS AND LICHENS.

ALSO OFFICIAL ORGAN
OF THE SULLIVANT MOSS CHAPTER OF THE AGASSIZ ASSOCIATION.

Subscription Price, $1.00 a year. 20c. a copy. Four issues 1898, 35c. Four issues 1899, 35c.
Together, eight issues, 50c. Four issues 1900, 50c. Four issues 1901, 50c. Four Vols. $1.50
Six issues 1902, $1.00. Six issues 1903, $1.00. Six issues 1904, $1.00.

*Short articles and notes on mosses solicited from all students of the mosses. Address manu-
script to A. J. Grout, Boys' High School, Brooklyn, N. Y. Address all inquiries and subscriptions to Mrs. Annie Morrill Smith, 78 Orange Street, Brooklyn, N. Y. For advertising space address Mrs. Smith. Check, except N. Y. City, MUST contain 10 cents extra
for Clearing House charges.*

THE SULLIVANT MOSS CHAPTER.

President, Mr. E. B. Chamberlain, Washington, D.C. Vice-President, Mrs. C.W. Harris,
125 St. Marks Avenue, Brooklyn, N. Y. Secretary, Miss Mary F. Miller. 1109 M Street,
Washington, D. C. Treasurer, Mrs. Smith, 78 Orange Street, Brooklyn, N. Y.
Dues $1.10 a year, this includes a subscription to THE BRYOLOGIST.
All interested in the study of Mosses, Hepatics, and Lichens by correspondence are
invited to join. Send dues direct to the Treasurer. For further information address the
Secretary.

BOTANICAL SUPPLIES
Everything for the Botanist

COLLECTING CASES—MOUNTING PAPER—MOUNTING CARDS
GENUS COVERS—TROWELS   SEND FOR CIRCULAR

GROUT

MOSSES WITH A HAND-LENS
NEW EDITION——JAN. 1905——$1.50 NET

MOSSES WITH A HAND-LENS AND MICROSCOPE

PARTS I. & II., $1.00 each net, postpaid.
Sample Pages on application.

O. T. LOUIS CO., 59 Fifth Ave., New York City

MOSSES OF THE SOUTHERN UNITED STATES
Sets of 51 Specimens (Nos. 1 to 51) may be had at $5.00 a set, on application to
DR. JOHN K. SMALL, BEDFORD PARK, NEW YORK CITY
JUST ISSUED!

FLORA OF THE SOUTHEASTERN UNITED STATES
BY JOHN K. SMALL, PH.D.

Being descriptions of the Flowering and Fern Plants growing naturally in North
Carolina, South Carolina, Georgia, Florida, Tennessee, Alabama, Mississippi, Arkansas,
Louisiana, The Indian Territory, and Oklahoma and Texas east of the one hundredth
Meridian, with analytical keys to the Orders, Families, Genera and Species.

Large octavo, pp. X + 1370   Price, $3.60

Subscriptions may be sent to DR. JOHN K. SMALL, BEDFORD PARK, NEW YORK CITY

WILLIAM STARLING SULLIVANT.

THE BRYOLOGIST.

Vol. VIII.   JANUARY, 1905.   No. 1.

WILLIAM STARLING SULLIVANT.
January 15, 1803—April 30, 1873.

A Biographical Sketch, adapted from that of Asa Gray, as given in the Supplement of the Icones Muscorum, 1874.

ANNIE MORRILL SMITH.

It is only fitting that the first place in this number of THE BRYOLOGIST should be given to a sketch of the life of the one for whom our Chapter is named, William Starling Sullivant. He was born at the little village of Franklinton, then a frontier settlement in the midst of the primitive forest, near the site of the present city of Columbus, Ohio. His father, a Virginian, and a man of marked character, was appointed by the government to survey the lands of that district of the "Northwest Territory" which became the central part of the now populous State of Ohio; and he early purchased a large tract of land, bordering on the Scioto River, near by, if not including, the locality which afterwards was fixed upon for the State Capitol. William was his oldest son. He received the rudiments of his classical education at the Ohio University at Athens, upon the opening of that institution, after a term in a Kentucky school; was transferred to Yale College where he was graduated in 1823. His father died that year and his services were demanded by the family to care for the estate, which was mainly in lands, mills, etc. To qualify for this he became a surveyor and practical engineer and took an active part in business till the latter part of his life. Mr. Sullivant was thrice married: his first wife was Jane Marshall, of Kentucky. She died within a year after marriage. His second was Eliza G. Wheeler, a lady of rare accomplishments, a zealous and acute bryologist, her husband's efficient associate in all his scientific work until her death of cholera, in 1850 or 1851. Her botanical services are commemorated in *Hypnum Sullivantæ* of Schimper, a moss then new to Ohio. His third wife, Caroline E. Sutton, survived him as well as children, grandchildren and great-grandchildren, all to inherit a stainless and honored name and to cherish a noble memory.

Mr. Sullivant was nearly thirty years old and already married, with his residence established in a suburban home surrounded by a rich flora, before his taste for such studies developed. He collected and carefully studied the plants of central Ohio, and made neat sketches of the minute parts of many of them, especially grasses and sedges, and began his correspondence with the leading botanists of the country, and in 1840 published "A Catalogue of Plants, Native or Naturalized, in the Vicinity of Columbus, Ohio," of sixty-three pages, to which he added a few pages of valuable notes. His only other publication in phanogamous botany is a short article on three new plants which he discovered in the district, contributed to the American Journal of Science and the Arts, in 1842. His further observations and notes were communicated to friends. As soon as the flowering plants of his district ceased to afford him novelty he turned to the mosses, in which he found abundant scientific occupation of a kind well suited to his bent for patient and close observation, scrupulous accuracy, and nice distinction and discrimination.

· His first publication in his chosen department was the "Musci Alleghaniensis," accompanied by the specimens themselves of Mosses and Hepaticæ collected in a botanical expedition through the Alleghany Mountains from Maryland to Georgia in the summer of 1843, Asa Gray being his companion. In 1846 Mr. Sullivant communicated to the American Academy the first part, and in 1849 the second part of his "Contributions to the Bryology and Hepaticology of North America," which appeared, one in the third, the other in the fourth volume (new series) of the Academy's Memoirs, each with five plates from the author's own admirable drawings. These plates were engraved at his own expense, and were generously given to the Academy. When the second edition of Gray's "Manual of the Botany of the Northern United States" was in preparation, Mr. Sullivant was asked to contribute to it a compendious account of the Musci and Hepaticæ of the region; which he did in the space of about one hundred pages, generously adding at his sole charge eight copper plates crowded with illustrations of the details of the genera, thus enhancing vastly the value of his friend's work and laying a foundation for the general study of bryology in the United States which then and thus began.

So excellent are these illustrations, both in plan and execution, that Schimper, then the leading bryologist of the Old World and a most competent judge since he has published hundreds of figures in his "Bryologia Europeæ," not only adopted the same plan in his Synopsis of the European Mosses but also the very figures themselves (a few of which, however, originally his own), whenever they would serve his purpose, as was the case with most of them. A separate edition was published of this portion of the Manual, under the title of "The Musci and Hepaticæ of the United States, east of the Mississippi River" (New York, 1856, imperial octavo) upon thick paper and with proof impressions directly from the copper plates. This exquisite volume was placed on sale at far less than cost, and copies are now of great rarity and value. It was with regret that the author of the Manual omitted this cryptogamic portion from the ensuing editions and only with the understanding that a separate "Species Muscorum" or Manual for the Mosses of the whole United States should replace it. This most needful work Mr. Sullivant was just about to prepare for the press, when death came to close his career. His work was, however, completed by his friends, Leo Lesquereux and Thomas P. James, and is the Manual of our daily use. For an account of his various Exsiccati reference can be made to the Icones Supplement Sketch by Asa Gray.

The "Icones Muscorum," however, is Mr. Sullivant's crowning work,

—3

as Prof. Gray says, and also the work with which we are most familiar. It consists, as the title indicates, of " Figures and Descriptions of most of those Mosses peculiar to Eastern North America which have not been heretofore figured," and forms an imperial octavo volume with one hundred and twenty-nine copper plates, published in 1864. The letterpress and plates are simply exquisite and wholly unrivalled, and the scientific character is acknowledged to be worthy of the setting. The second volume was in course of preparation at the time of Mr. Sullivant's death, but the material was found to be mostly in notes on herbarium sheets, etc., and the work of editing was undertaken by Leo Lesquereux who alone was in a position to complete it. This was done as a labor of love for his friend, and though pressure was brought to bear to have the name of Leo Lesquereux appear on the title page, he would not consent, and it appears as the final work of Sullivant, though the preface acknowledged this indebtedness to Lesquereux.

In accordance with his wishes all his bryological books and his exceedingly rich and important collections and preparations of mosses were consigned to the Gray Herbarium of Harvard University with a view to their safe keeping and long continued usefulness. The remainder of his botanical library, his choice microscopes, and other collections went to the State Scientific and Agricultural College established at the time of his death at Columbus, and to the Starling Medical College, founded by his uncle and of which he was himself the senior trustee.

Mr. Sullivant was chosen into the American Academy in 1845; received the honorary degree of Doctor of Laws from Gambier College in his native State, was an associate of the principal scientific societies of this country and of several in Europe. His oldest botanical associates long ago enjoyed the pleasure of bestowing the name SULLIVANTIA OHIONIS upon a very rare plant, a Saxifrage, which he himself discovered in his native State on the secluded banks of a tributary of the river which flows by the place where he was born and where his remains now repose.

SPORE DISTRIBUTION IN BUXBAUMIA.

A. J. GROUT.

Mr. Dixon in his Handbook of British Mosses states that *Buxbaumia aphylla* scatters its spores by the rupture of the capsule walls. Schimper in the Bryologia Europea states that the tube of the peristome is so narrow that the spores cannot pass out after the capsule dies and the peristome becomes twisted.

The peristome of Buxbaumia is so perfectly developed that it has not seemed probable to me that it could be a useless organ, and for several years I have been trying to get fresh specimens just at the time of complete maturity and before the spores had been shed. Early last June Mr. Walter Gerritson sent me in some specimens which were in just the right condition and when the capsules were lightly tapped with a pencil the spores were projected as far and as freely as in *Webera sessilis*. After dehiscence the capsules partially collapse so that undoubtedly some of the spores do escape by the breaking of the capsule walls but that this is the main reliance of the species I do not for a moment believe.

Prof. Goebel says that the breaking of the outer walls of the capsule of *B. indusiata* renders it easier for the raindrops to force out the spores (by reason of the lessened resistance of the capsule wall to the impact of the drops) so that he evidently believes that the peristome of Buxbaumia is functional.

NOTES ON NOMENCLATURE IV.—THE GENUS NECKERA HEDW.

By Elizabeth G. Britton.

There have been three genera named for Noel J. Necker (1729–1793):

Neckeria Scopoli Int. 313. 1777 equals *Capnoides* (Papaveraceæ).

Neckeria Hedw. Fund. 2:93. 1782 equals *Neckera* Hedw. (Neckeraceæ).

Neckeria Ait. Gmel. Syst. 3:316. 1791. equals *Pollichia* (Caryophyllaceæ).

The first genus named for him is not in use at present, being antedated by *Capnoides* Adans, but as there are one hundred and ten species in this genus, if it be divided, *Neckeria* of Scopoli would have precedence over *Neckeria* Hedw. Recognizing this fact Mr. S. C. Stuntz published in 1900 a Revision of the North American Species of *Neckera* Hedw., taking up the generic name of *Eleutera* Beauv. (1805). This name is antedated by *Rhystophyllum* Ehrh. (1780–1789) which was founded on *Hypnum crispum* L. (1753), which in turn was based on the descriptions and illustrations given by Dillenius (1741) and Robert Morrison (1699), both of which are unmistakably referable to the genus *Neckera* as at present understood.

As originally founded by Hedwig (1782) his genus *Neckeria* was described simply as having a double peristome, and included *Hypnum crispum, curtipendulum, viticulosum* and *sericeum* which have been referred to *Neckera, Antitrichia, Anomodon* and *Homalothecium*. The type species is the same as in *Rhystophyllum*, but that Hedwig did not understand the genus at all in its modern restricted sense is shown by his treatment of *Neckera*, in his Muscorum Frondosorum, ten years later, when he figured nine species which have since been referred by other authors to *Pilotrichella, Pilotrichum, Pterobryum* and *Cylindrothecium*, including four species of *Neckera*. Furthermore, Hedwig included one species of *Neckera* in his genus *Leskia* (1782) which was also a mixture, including *Pylaisea, Anomodon, Eurhynchium* and *Leskea*. In fact it will be found that the descriptions given by Morrison and Dillenius, and quoted by Linnaeus, are more lucid and applicable to the genus, than those given by Hedwig, and as Ehrhart's genus *Rhystophyllum* is monotypic, being based on one Linnean species with two illustrations, and founded on a specimen issued in a set of Exsiccatae, there is no question as to his meaning or the application of the name, seeing that its derivation from two Greek words meaning Wrinkled-Leaved, indicates one of the most noticeable characters of the genus as limited in modern times.

It may be of interest to notice the variety of species and genera which have been included under *Neckera* up to 1850 when it was reduced to its natural limits by Schimper. Bridel (1801) described fifteen species, adding some belonging to *Cryphaea* and *Climacium*, to those previously included by Hedwig. In 1819 Bridel divided the genus into two sections, including the species of modern authors in his section *Distichia*, and placing species of *Anomodon* and *Cylindrothecium* under *Neckera*. In the Bryologia Universalis (1827) he added another section to the genus and kept the species of *Neckera* under *Distichia*. Carl Müller took up *Distichia* as a genus and described an African species in 1890.

The genus *Eleutera* was founded by Beauvois (1805) as a substitute for *Neckera* because he did not believe in naming genera after persons so he substituted a name applied to Diana! He listed seven species belonging to *Anomodon*, *Antitrichia* and *Neckera*, of which five had been included in *Neckera* by Hedwig, adding two species of *Cylindrothecium*.

Schimper, in the Bryologia Europaea, 1850, figured and described five species of *Neckera: pennata, oligocarpa, pumila, crispa* and *complanata*, thus bringing the genus into its natural limits, and most subsequent authors have followed him.

But for comprehensiveness and amplification of the genus *Neckera*, Carl Müller exceeded all others, for in 1851, a year after the publication of the fascicle on *Neckera* in the Bryologia Europaea, he described one hundred and fifty-two species with nine sections and thirteen subsections including, according to his own statement, the following genera: *Braunia, Hedwigidium, Entodon, Dichelyma, Leucodon, Asterodontium, Antitrichia, Sclerodontium, Hedwigia, Harrisonia, Leptodon, Lasia, Isothecium, Rhystophyllum, Climacium, Pterigynandrum, Leptohymenium, Pilotrichum* and parts of *Leskea, Hypnum,* and *Fontinalis*. It is one of the subsections, *Cryphaeadelphus*, which M. Cardot has recently raised to generic rank to replace *Brachelyma* Sch. If all the old sectional and subsectional names which antedate generic names are to be hunted up there will be no end to the changes and the work necessary to get questions of priority correctly determined!

Jaeger in the Adumbratio (1875--76) recognized one hundred and four species and two sections of the genus, *Paraphysanthus* Spruce, and *Rhystophyllum* Ehrh., and included in the latter five species recognized by Schimper, adding *Menziesii* and *Douglasii*. Paris, in the Index, recognizes one hundred and fifty-eight species of *Neckera*, of which fifty are American and twenty-five are North American and West Indian.

The validity of Ehrhart's genera is being recognized, and Brotherus in the Pflanzenfamilien adopts *Georgia, Catharinea,* and *Webera* and relegates to synonymy *Tetraphis, Atrichum, Webera* Hedw. and *Diphyscium* Ehrh. and we believe that *Rhystophyllum* Ehrh. also has valid claims.

Rhystophyllum Ehrh. Beitr, 149, 1789. Crypt. Exsic. No. 97. 1780.

Neckeria Hedw. Fund. 93. 1782 in part.

Leskea Hedw. Fund. 93. 1782 in part.

Neckera Hedw. Musc. Frond. **3**:48. 1792 in part.
Eleutera Beauv. Prod. 35. 1805 in part.
Neckera Sch. Br. Eu. fasc. 44-45. 1850.
Type species *Hypnum crispum* L. Sp. pl. **2**:1124. 1753.
The following species are at present known in North America:

1. { RHYSTOPHYLLUM DOUGLASII (Hook).
{ *Neckera Douglasii* Hooker, Bot. Misc. **1**:131. pl. 35. 1830.
2. { RHYSTOPHYLLUM PENNATUM (L.)
{ *Fontinalis pennata* L. Sp. Pl. 1371. 1763.
{ *Neckera pennata* Hedw. Musc. Frond. **3**:47. pl. 19. 1792.
3. { RHYSTOPHYLLUM OLIGOCARPUM (Bruch.)
{ *Neckera oligocarpa* Bruch, Mscr. in Hartm. Skand. Fl. 338. 1849.
4. { RHYSTOPHYLLUM MENZIESII (Hook.)
{ *Neckera Menziesii* Hook. in Drum. Musc. Bor. Am. (Ed. 1.)
5. { RHYSTOPHYLLUM ONITHOPODIOIDES (Scop.)
{ *Hypnum ornithopodioides* Scop. Fl. Carn. 164. 1760.
{ *Neckera complanata* Hub. Muscol. Germ. 576. 1832.
6. { RHYSTOPHYLLUM DISTICHUM (Sw.)
{ *Fontinalis distichum* Sw. Pr. Fl. Ind. Occ. 138. 1788.
{ *Neckera distichum* Hedw. Musc. Frond. **3**:53. pl. 22. 1792.
7. { RHYSTOPHYLLUM JAMAICENSIS (Gmel.)
{ *Hypnum Jamaicensis* Gmel. L. Syst. Nat. 1341. 1791.
{ *Neckera undulata* Hedw. Musc. Frond. **3**:51? pl. 21. 1792.

BOOK NOTICE—THE TEACHING OF BIOLOGY, BY F. E. LLOYD AND M. A. BIGELOW.

A. J. GROUT.

It is not often that THE BRYOLOGIST feels called upon to review books on other subjects than those to which it professes to devote itself. However, so many of our readers are also teachers that we feel they will thank us for calling their attention to this book which is not merely excellent in theory, but is full of practical hints and suggestions as to material and method.

No teacher of biology or nature study can read this book without being helped. It is almost needless to say that Prof. Lloyd treats of Botany and Prof. Bigelow of Zoölogy.

It is published by Longmans, Green & Co. (in the American Teachers' Series), New York, 8vo., $1.50.

MUSCI BOREALI-AMERICANI BY PROF. J. M. HOLZINGER.

A. J. GROUT.

Fascicle 5, numbers 101-125 of Prof. Holzinger's Musci Acrocarpi Boreali-Americani has just come to hand. Some of the more interesting species are : Mnium glabrescens Kindb., M. venustum Mitt., Bryum coronatum Schwaegr., B. Sawyeri R. & C., B. cirrhatum Hoppe., Orthotrichum pulchellum Brunton and its variety leucodon Vent., Funaria Americana Lindb., Webera proligera (Lindb.) Kindb., Scouleria aquatica Hook., Fissidens rufulus B. & S., and Dicranodontium longirostre (Web. & Mohr.) B. & S.

REVIEW OF DR. WARNSTORF'S PAPER ON EUROPEAN HARPIDIA.

JOHN M. HOLZINGER.

This paper is published in the Beiheft zum Botanischen Centralblatt, Band XIII, Heft 4, 1903. It is accompanied by two plates.

In dealing with the literature of the group the author naturally reviews the work of previous authors who have made it a specialty. Carl Mueller was the first to treat this as a subsection of Hypnum, under the name of *Drepanocladus* (1851). In 1856 Sullivant gave the name *Harpidium* to essentially the same group, with some omissions, and for forty years Sullivant's name has been in use, in spite of the priority of Mueller's name. Since 1885, however, there has been established a genus of lichens, *Harpidium* Körben. So that now a double ground exists for holding to Mueller's older *Drepanocladus*. Limpricht, in his Laubmoose, has reinstated the name, and Warnstorf very properly has taken like ground.

The authors most interested in this group were Schimper (1876), Sanio (1891), Renauld (1890), v. Klingraeff (1893), and Limpricht (1900). The least practical, most mechanical treatment appears to have been that of Sanio. Both Renauld and Limpricht are far more logical in disposing of the multitude of forms in this polymorphic group. A number of species retained by them are however reduced by Warnstorf, and his discussion of these species is quite instructive.

Following is his synopsis of European species of *Drepanocladus*:

I. INTEGROFOLIA—

    A. GROUP KNEIFFII:
        D. Kneiffii (Sch.)
        D. polycarpus (Bland).
        D. pseudofluitans (Sanio).
        D. simplicissimus (W. Warnst.)

    B. GROUP ADUNCUS:
    Crassicostata—
        D. capillifolius (Warnst.)
        D. aduncus (Hedw.)
        D. Sendtneri (Sch.)
    Tenuicostata.
        D. lycopodioides (Schwaegr.)
        D. latifolius (Lindb. & Arnell).
        D. brevifolius (Lindb.)
        D. subaduncus (Warnst.)

    C. GROUP INTERMEDIUS:
        D. vernicosus (Lindb.)
        D. intermedius (Lindb.)
        D. revolvens (Sw.)
        D. latinervus (Arnell).

II. SERRATIFOLIA—

    D. GROUP UNCINATUS:
        D. uncinatus (Hedw.)

    E. GROUP EXANNULATUS:
        D. Rotae (DeNot).
        D. pseudorufescens (Warnst.)
        D. fluitans (L.)
        D. exannulatus (Guemb.)

The author's interruption in this work, in order to elaborate his local Mossflora, has prevented his going into much greater detail and after spreading out his general plan he describes minutely and discusses the range of forms of only the eight species which appear to occur in his restricted area, vis.: *D. pseudofluitans;* D. *simplicissimus,* with three new varieties: *D. capilifolius,* with five varieties, one new; *D. lycopodioides; D. latifolius; D. brevifolius: D. uncinatus,* with eight varieties, one of them new; *D. subaduncus.*

In a foot note under *D. uncinatus plumulosus* the author considers *Hypnum symmetricum* Ren. & Card., a form of this plumulose variety. No other American forms are referred to. While the author does not complete his task in the sense of his first intention, his treatment of this difficult section of Hypnum, based upon a comprehensive study of all the important Harpidium collections of Europe, commands both respectful attention and interest in the sound common sense judgment shown. And one of these days when from among our own bryologists one shall take up our American Harpidia for a general review, the way pointed out by this author will be of the greatest possible value.   Winona, Minn.

NEW OR UNRECORDED MOSSES OF NORTH AMERICA.

BY J. CARDOT AND I. THÉRIOT.

Translated and condensed from The Botanical Gazette, May, 1904.

DESCRIPTIONS OF NEW SPECIES GIVEN IN FULL.

PHASCUM HYALINOTRICHUM Card. & Thér.

Plants small, budlike, solitary or clustered, light green. Leaves imbricated, the lower smaller, the upper larger, median and upper ovate, o.8-1 mm. long by o.6 mm. wide, concave, entire or subentire, acuminate, margins plane or slightly reflexed below: costa narrow, 24μ wide in the middle of the leaf, attenuate below, excurrent into a hairlike flexuous hyaline point, $\frac{1}{3}$-$\frac{1}{2}$ the length of the leaf. Areolation lax, not papillose not very chlorophyllose, hyaline above, median cells irregular, quadrate, short-rectangular or subhexagonal, 18-30μ by 12-18μ, somewhat incrassate, the lower larger and thinner walled, the upper longer, more incrassate. Seta very short, o.2 mm. long, geniculate; capsule immersed, globose, apiculate, o.7 mm. in diameter; calyptra cucullate, covering half the capsule. Ripe spores unknown. Plate XVI.

California: Soldiers' Home, Los Angeles Co. (Dr. Hasse, 1902: herb.
C. F. Baker).

A quite peculiar species, very distinct by its habit, which recalls that of
an Acaulon, its puliform excurrent nerve, and its loose smooth areolation, a
little chlorophyllose below and hyaline above.

PLEURIDIUM BAKERI Card. and Thér.

Plants short, loosely caespitose, yellowish green. Stem 2-4 mm. long,
erect, simple. Leaves erect, the lower minute, distant, the upper longer,
lanceolate-subulate, acute, entire, rarely subdenticulate at apex, subcanalic-
ulate by the inflexed margins, 1.3 by 0.35 mm.: perichaetial leaves twice as
long, gradually subulate; costa broad, 80-100μ, percurrent, somewhat narrower in the perichaetial leaves, lower cells pellucid, subrectangular, 24μ by
12μ, median and upper narrower, 29-30μ by 5μ, opaque, incrassate. Seta
erect, short, 0.4 mm. long; capsule immersed, ovate, somewhat gibbous,
obtusely apiculate, smooth, orange when mature, 1 mm. by 0.6 mm. Calyptra cucullate, covering ⅓-½ the capsule. Spores subglobose, minutely
papillose, 24-30μ in diameter. Seemingly dioicous, antheridial buds
unknown. Plate XVI.

California: On ground in old pastures, foothills near Stanford University (C. F. Baker, 1902).

VAR. ELONGATUM Card. & Thér.

Differs from the typical form in the longer stems and longer and more
flexuous upper leaves.

California: On wet clay soil, foothills near Palo Alto (C. B. Baker, 1902).

Distinguished from *P. subulatum* Br. Eur. by the shorter and less
finely subulate leaves, with a broader costa. The later character also separates our species from *P. Bolanderi* C. Muell., which, besides, has the
leaves distinctly denticulate on the margins from the middle upward. *P.
Ravenelli* Aust., of which we have seen no authentic specimen, according
to the description has carinate leaves, excurrent costa, and synoicous inflorescence. If the inflorescence of *P. Bakeri* is, as we think, really dioicous,
this character would distinguish it from all the other North American and
European species of Pleuridium.

DICRANELLA CURVATA Sch. var. MISSOURICA Card. & Thér.

Differs from the type in the less distinctly striate capsule and the
broader and shorter leaf cells.

Missouri: Seligman, on ground (B. F. Bush, 1898).

DICRANUM ALATUM (Barnes) Card. & Thér.

Dicranum Bonjeani DeNot. var. *alatum* Barnes.

Illinois: Chicago (Dr. J. Röll, 1888). Wisconsin: Madison (Cheney
and True). W. Minnesota: Cedar Lake, near Montevideo, Chippewa Co.
(J. M. Holzinger, 1901).

The nerve bearing on the back two high, serrate lamellae, and the
shorter, thinner-walled cells of the areolation, seem characters of sufficient
value to separate this moss from *D. Bonjeani*. Plate XVII.

FISSIDENS BUSHII Card. & Thér.

Fissidens subbasilaris var. *Bushii* Card. & Thér.

Missouri: Eagle Rock, on gravelly ground (B. F. Bush, 1897). Texas: (Elsa Baumann; herb. Dr. Zickendrath).

At first we considered this moss as a variety of *F. subbasilaris* Hedw. but further observations led us to a different conclusion, and now we think it preferable, on account of the monoicous inflorescence and the costa reaching the apex, to place it near *F. taxifolius* Hedw., of which it may be a subspecies, characterized by the nearly twice smaller size, the obtuse abruptly apiculate leaves with the dorsal wing not undulate at base and with smaller cells. From *F. subbasilaris* it is easily distinguished by the inflorescence and the costa reaching the apex. Plate XVII.

FISSIDENS PUSILLUS Wils. var. BREVIFOLIUS Card. & Thér.

Differs from the type in the much shorter and more abruply and shortly acuminate leaves of the sterile innovations.

California: Soldiers' Home, Los Angeles Co. (Dr. Hasse, 1902; herb. C. F. Baker).

POTTIA NEVADENSIS Card. & Thér.

Monoicous? green: caespitose. Stems erect, short, 1-2 mm. long. Leaves erect, ovate, concave, median 1 by 0.6 mm., the lower smaller, all smooth with entire margins somewhat revolute on one side at least. Costa narrow, abruptly excurrent into a subpiliform apex, median cells hexagonal or rhomboidal, not very chlorophyllose, about 18μ by 15μ, the upper somewhat smaller, the basal elongated rectangular 40μ by 18μ, all entirely smooth. Perichaetial leaves much larger and lighter colored, broader, strongly concave, reaching 2 mm. in length. Seta light colored, flexuous, 10-15 mm. long. Capsule 1.5-2 mm. by 0.75 mm., erect, oblong, attenuate at base, truncate at mouth, pale, without peristome. Operculum depressed, long and slenderly beaked, 1.2 . mm. long. Columella adhering somewhat. Spores densely papillose, 24-30μ in diameter. Plate XVIII.

Nevada: Kings Cañon, near Carson, on ground about willow thickets (C. F. Baker, 1902).

This species seems very distinct from any other European or North American Pottia.

DIDYMODON TOPHACEUS Jur. var. DECURRENS Card. & Thér.

Similar to var. *elatus*. Leaves remote, recurved when moist, long decurrent, strongly papillose on the back, costa rough.

Texas: Shovel Mt., Burnet Co. (Rev. Franciscus Ebeling: herb. Dr. E. Zickendrath).

DESMATODON BUSHII Card. & Thér.

Plants loosely caespitose, olive-green. Stems erect, 5-15 mm. long simple or divided. Leaves crispate when dry, when moist erect-spreading 1-1.5 mm. by 0.4 mm., oblong-lanceolate, mostly obtuse, costa short-excurrent, mucronate, margins entire, longly revolute, lower cells rectangular, hyaline or somewhat yellowish, 35-40μ by 9μ, the rest roundish-quadrate, strongly papillose, opaque, 7-8μ in diameter. Costa 60μ thick at base. Perichaetial leaves larger, hyaline in the lower $\frac{1}{3}$-$\frac{1}{2}$, more narrowly acuminate, margins about as much revolute above. Seta pale red, about 10 mm. long. Capsule erect, subcylindric, with operculum about 2 by 0.44 mm., operculum conic, rather short, .05 mm. long. Calyptra covering $\frac{1}{2}$ the capsule. Peristome 0.35 mm. high, purple, basal membrane short, teeth little twisted, divided to the base into two filiform divisions, papillose, divisions usually united below by two or three trabeculæ. Spores smooth, variable, globose or elliptic, 20-30μ in diameter. Seemingly dioicous (antheridial buds unknown). Plate XVII.

Missouri: Courtney, on wet rocks (B. F. Bush, 1898).

Much resembling *Barbula unguiculata* Hedw. by the habit and the shape and areolation of the leaves, but readily distinguished from it by the peculiar structure of the peristome, which is hardly twisted, and by the spores twice larger. C. Mueller describes a *B. cancellata*, the peristome of which according to the description must have a similar structure to that of our *D. Bushii*, but the teeth are smooth (likely twisted) and the lid is as long as the capsule.

DESMATODON SYSTILIOIDES Ren. & Card.

This is not a *Desmatodon* but a new *Pottia* of the group *Heimii*.

BARBULA MACROTRICHA Card. & Thér.

Tufts compact, hoary above. Stems short, 1-2 mm. long. Leaves erect when moist, appressed into a bud-like shape, twisted when dry, 1-1.5 mm., by 0.6-0.8 mm. ovate or short subspatulate, apex broadly obtuse, often emarginate especially in the lower leaves, broken when old, margins entire, plane at base, above this revolute to near the apex. Costa narrow, 4-8μ thick, excurrent into a smooth hyaline hair which is short in the lower leaves but equalling the rest of the leaf in the upper leaves. Lower leaf cells quadrate or short rectangular, 18μ broad, hyaline or slightly chlorophyllose, nearly smooth, the other cells smaller, about 12μ broad, quadrate or subrotund, strongly chlorophyllose and highly papillose, therefore indistinct. Seta reddish at base, above pale, 6-8 mm. long, Capsule erect, subcylindric, somewhat curved, 2.5 mm. long, including the elongated conic operculum. No mature capsules known. Seemingly dioicous as antherial plants were not seen on fruiting plants. Plate XVIII.

California: Soldier's Home, Los Angeles Co. (Dr. Hasse, 1902; herb. C. F. Baker).

In general appearance resembling the smallest forms of *B. muralis* Timm., from which it differers by the smaller size, the short pedicel, the much shorter leaves obtuse or emarginate and finally eroded at the apex, the hair of the upper leaves as long as or even longer than the lamina, etc.

(To be continued.)

SULLIVANT MOSS CHAPTER ANNUAL REPORTS.

REPORT OF JUDGE OF ELECTIONS.

MISS MARY F. MILLER,   December 1, 1904.
   Secretary, Sullivant Moss Chapter.

The following report of the election of officers of the Chapter for the year 1905 is respectfully submitted:

For President—Mr. Edward B. Chamberlain 20
"   Vice-Pres.—Mrs. Carolyn W. Harris................ 21
"   Secretary—Miss Mary F. Miller.................... 20
"   Treasurer—Mrs. Annie Morrill Smith.............. 21

The above candidates are elected.

                                        AGNES CHASE,
                                        Judge of Elections.

REPORT OF THE TREASURER.

The following statement for the year beginning December 1, 1903, and ending December 1, 1904, is respectfully submitted:

RECEIPTS:		DISBURSEMENTS:	
By cash in hand Dec. 1, 1903..	$15 57	To the BRYOLOGIST..........	$96 30
" " to balance from Miss		" Express..................	2 85
Wheeler...	1 65	" Postage and Stationery..	9 53
" " sale of Herbarium			
Cases.	1 50		$108 68
" Dues from Members....	118 88	Cash in hand Dec. 1, 1904...	28 92
	$137 60		$137 60

ANNIE MORRILL SMITH, Treasurer.

REPORT OF THE SECRETARY.

The records of the Sullivant Moss Chapter for 1904 show a gratifying continuance of its usual prosperity. There are now one hundred and thirty-eight members, twenty-two new ones having been added during the year. We have lost five members; four have withdrawn, and I regret to record the death of Mrs. Emilia C. Anthony. Mrs. Anthony always displayed an active interest in bryology, and contributed generously to the Chapter Herbarium.

There are at present nearly one thousand three hundred specimens in the Herbarium of the Moss Department of the Chapter, representing three hundred and eighty-one species and varieties and one hundred and nine genera, one hundred and fifty-nine new specimens and thirty-six new species having been added this year. Among the larger contributions to the Herbarium are over eighty mosses from Florida, North Carolina and Massachusetts, donated by Miss Abby M. Small; and a number of Florida mosses, donated by Mr. Severin Rapp. There have also been valuable con-

tributions from other members. During 1904, seventy-nine specimens of
mosses, hepatics and lichens have been offered in the BRYOLOGIST. It is
hoped that members will offer more hepatics during the coming year.

A great many mosses have been sent in for determination, and the
experts have ever responded cheerfully and willingly to the call for assistance in naming them. Appreciation should also be expressed of the hearty
co-operation of the collectors throughout the country, for it is owing to their
untiring energy in collecting and distributing specimens that the "offer-
ing" department of the Chapter is possible.

Respectfully submitted,

MARY F. MILLER,
Secretary.

OFFERINGS.

(To Chapter Members only. For postage.)

Mrs. R. H. Carter, 37 Church street, Laconia, N. H. *Rhacomitrium acicu-
lare* (L.) Brid., c.fr.; *Rhynchostegium rusciforme* (Neck.) B. & S., c.fr.,
Hypnum ochraceum Turn. st.; *H. palustre* Hedw., c.fr. Collected in
Gilford, N. H.

Mr. N. L. T. Nelson, 3968 Laclede Ave., St. Louis, Mo. *Gymnostomum
curvirostre* (Ehrh.) Hedw., c.fr.; *Ditrichum pallidum* (Schreb.)
Hampe., c.fr. Collected in Missouri.

Mr. Edward B. Chamberlain, 1830 Jefferson Place, Washington, D. C.
Hylocomium proliferum (L.) Lindb., c.fr. Collected in Maine.

Mr. Severin Rapp, Sanford, Orange Co., Florida. *Leucobryum sediforme*
(Müll.) Hedw., c.fr.; *Funaria hygrometrica* (L.) Sibth. var. *lutea*, c.fr.
Collected in Sanford.

Mrs. Sarah B. Hadley, South Canterbury, Conn. *Dichelyma capillaceum*
B. & S., c.fr,; *Leucobryum glaucum* (L.) Sch., c.fr. Collected in South
Canterbury.

Prof. W. W. Stockberger, Bureau Plant Industry, Washington, D. C.
Mnium punctatum Hedw., c.fr.; *Dicranum viride* Schimp., st. Collected in the White Mountains, N. H.

Mr. Charles C. Plitt, 1706 Hanover street, Baltimore, Md. *Frullania Asagrayana* Mont. Collected near Baltimore.

Mr. G. K. Merrill, 564 Main street, Rockland, Maine. *Cetraria nivalis* Ach.
Collected on Mt. Washington, N. H.

LIST OF SULLIVANT CHAPTER MEMBERS

Adams, Miss Carrie E.................................Hinsdale, N. H.
Adams, Mr. F. M ....337 Greene Ave., Brooklyn, N. Y.
Ainslie, Mr. Charles N..............First National Bank, Rochester, Minn.
Ames, Mr. Oakes...........Ames Botanical Library, North Easton, Mass.
Anderson, Mr. John A...............   ....High School, Dubuque, Iowa.
Badè, Dr. Wm. F.................University of California, Berkeley, Cal.
Bailey, Dr. J. W...Seattle, Wash.
Bailey, Miss Harriet B.............830 Amsterdam Ave., New York City.
Barbour, Mr. Wm. C ....Sayre, Pa.
Barnes, Prof. Charles R .....Dept. Botany, Univ. of Chicago, Chicago, Ill.
Best, Dr. George N...   ....Rosemont, N. J.
Bonser, Prof. Thomas A...   ....Spokane, Wash.
Brenckle, Dr. J. F. .... Box 204, Kulm, North Dakota.
Britton, Mrs. Elizabeth G.......N. Y. Botanical Garden, Bronx Park, N. Y.
Brown, Mr. Edgar....Division of Botany, Dept. Agric., Washington, D. C.
Browne, Mrs. A. F................... ....Mahone Bay, Nova Scotia.
Bruce, Mr. C. Stanley........................ ...Shelburne, Nova Scotia.
Bryant, Miss E. B....................32 Reedsdale Street, Allston, Mass.
Carr. Miss C. M......................................South Sudbury, Mass.
Carter, Mrs. R. H................... ....37 Church St., Laconia, N. H.
Chamberlain, Mr. Edward B........1830 Jefferson Place, Washington, D. C.
Chapin, Mrs. Louis N................. ...11 East 32nd Street, N. Y. City.
Chase, Mrs. Agnes...............59 Florida Ave., N.W.,Washington. D. C.
Chase, Mr. Virginius H........................Wady Petra, Stark Co., Ill.
Cheney, Prof. L. S ....Barron, Barron Co., Wis.
Chatterton, Mr. F. W... ....227 Townsend Ave., New Haven, Conn.
Choate, Miss Agnes D..................3400 Morgan Street, St. Louis, Mo.
Clapp, Mrs. J. B.............. ... .. 52 Hartford Street, Dorchester, Mass.
Clark, Mr. H. S..................,..........16 Lineten Place, Hartford, Conn.
Clarke, Mrs. Sarah L...... ....1 West 81st Street, N. Y. City.
Clarke, Miss Cora H..................91 Mt. Vernon Street, Boston, Mass.
Coffin, Miss Mary F................115 Newtonville Ave., Newton, Mass.
Collins, Mr. J. Franklin.................468 Hope Street, Providence, R. I.
Coomes, Mrs. Laura M..................Queens, Queens Co., N. Y. City.
Craig, Mr. T....................1013 Sherbrooke Street, Montreal, Canada.
Cresson, Mr. Ezra T. Jr........................Box 248, Philadelphia, Pa.
Crockett, Miss Alice L..................................... Camden, Maine.
Cummings, Prof. Clara E..............Wellesley College, Wellesley, Mass.
Curtis, Mrs. Elizabeth B....................Box 47, Hendersonville, N. C.
Cushman, Miss H. Mary...............300 North Fifth Street, Reading, Pa.
Dacy, Miss Alice E.................... 28 Ward Street, South Boston, Mass.
Demetrio, Rev. Charles H.................... ...Emma, Salina Co., Mo.
Doran, Miss Genevieve.............13 Washington Ave., Waltham, Mass.
Dunham, Mrs. Horace C. M...........53 Maple Street, Aurburndale, Mass.
Dupret, Mr. H.................Seminary of Philosophy, Montreal, Canada.

Eaton, Mr. Alvah H.............................. ....Seabrook, N. H.
Eby, Mrs. Amelia F.................141 North Duke Street, Lancaster, Pa.
Edwards, Prof. Arthur M.............. .....423 Fourth Ave., Newark, N. J.
Evans, Dr. Alexander W..... ....2 Hillhouse Ave., New Haven, Conn.
Fink, Prof. Bruce.................... ....Grinnell, Iowa.
Fletcher, Mr. S. W..Pepperell, Mass.
Frye, Prof. T. C.. ... .State University, Seattle, Wash.
Gerritson, Mr. Walter .. ..66 Robins Street, Waltham, Mass.
Gilbert, Mr. B. D ....Clayville, Oneida Co., N. Y.
Gilman, Mr. Charles W....................Palisades, Rockland Co., N. Y.
Gilson, Miss Helen S ....50 Williams Street, Rutland, Vermont.
Graves, Mr. James A .... Susquehanna, Pa.
Greenalch, Mr. Wallace...............54 North Pine Street, Albany, N. Y.
Greever, Mr. C. O......... ....1345 East Ninth St., Des Moines, Iowa.
Grout, Dr. A. J...........................360 Lenox Road, Brooklyn, N. Y.
Hadley, Mrs. Sarah B ....South Canterbury, Conn.
Harris, Mrs. Carolyn W.............. 125 St. Mark's Ave., Brooklyn, N. Y.
Harris, Mr. Wilson P ....987 West Delevan Ave., Buffalo, N. Y.
Haughwout, Miss Mary R.......................Patton, Cambria Co., Pa.
Haynes, Miss Caroline C...................16 East 36th Street., N. Y. City.
Haydock, Mr. Wm. E.............. ...1328 Chestnut St., Philadelphia, Pa.
Hill. Mr. Albert J............... .....New Westminster, British Columbia.
Hill, Mr. E. J.... ....7100 Eggleston Ave., Chicago, Ill.
Holzinger, Prof. John M ....Winona, Minn.
Horton, Mrs. Francis B ....13 Brook Street, Brattleboro, Vt.
House, Mr. Homer D ....U. S. National Museum, Washington, D. C.
Huntington, Mr. J. Warren ....Amesbury, Mass.
Hurlbut, Mrs. R. H...... .... South Sudbury, Mass.
Jackson, Mr. Joseph......... ....16 Woodland Street, Worcester, Mass.
Jennings, Mr. Otto E.. ...419 Craft Ave., Carnegie Museum, Pittsburg, Pa.
Joline, Mrs. A. H ....1 West 72nd St., N. Y. City.
Jones, Mr. Wm..... ....Box 120, Lewiston, Fulton Co,, Ill.
Jump, Mrs. Harvey D.................................. ....Sayre, Pa.
Kennedy, Dr. George G. ....Readville, Mass.
Kendall, Miss Alice C. ....Bird's Oak, Auburndale, Mass.
Klem, Miss Mary J...................1808½ Lafayette Ave., St. Louis, Mo.
Krout, Prof. A. F. K...... ....Glenolden, Delaware Co., Pa.
Lamprey, Mrs. E. S .. ..2 Guild Street, Concord, N. H.
Lippincott, Mr. Charles D.... ....Swedesboro, New Jersey.
Lorenz, Miss Annie........ ....96 Garden Street, Hartford, Conn.
Lowe, Mrs. Josephine D....................Noroton, Fairfield Co., Conn.
Marshall, Miss M. A......... ....Still River, Worcester Co., Mass.
Martens, Mr. J. W. Jr. ....Shrub Oak, Westchester Co., N. Y.
Mathews, Miss Caroline.... ....Waterville, Maine.
Maxon, Mr. Wm. R ....U. S. National Museum, Washington, D. C.
McConnell, Mrs. S. D........... ....781 Madison Avenue, N. Y. City.

McDonald, Mr. Frank E.. ... .417 California Ave., Peoria, Ill.
Merrill, Mr. G. K......... ....564 Main Street, Rockland, Maine.
Metcalf, Mrs. Rest E...Hinsdale, N. H.
Miller, Miss Mary F...... ....1109 M St., N. W.,Washington, D. C.
Miller, Mr. Robert K............ .....14 East Pleasant St., Baltimore, Md.
Mirick, Miss Nellie...................28 East Walnut Street, Oneida, N. Y.
Murray, Miss Elsie........................... ....Athens, Pa.
Naylor, Prof. J. P .. ..Greencastle, Ind.
Nelson, Mr. N. L. T....................3968 Laclede Ave., St. Louis, Mo.
Newman, Rev. S. M ....Cor. 10th & G Sts., N. W., Washington, D. C.
O'Connor, Mrs. J. T....................................Garden City, N. Y.
Oleson, Mr. O. M..Fort Dodge, Iowa.
Palmer, Mrs. Rebecca L............. ...615 Putnam Ave., Brooklyn, N. Y.
Perrine, Miss Lura L.........State Normal School,Valley City, N. Dakota.
Plitt, Mr. Charles C..................1706 Hanover Street, Baltimore, Md.
Pratt, Miss Henrietta A.................63 Central Street,Waltham, Mass.
Pollard, Mr. Charles L..................286 Pine Street, Springfield, Mass.
Puffer, Mrs. James J.... .... Box 39, Sudbury, Mass.
Rapp, Mr. Severin.........................Sanford, Orange Co., Florida.
Rau, Mr. Eugene A..Bethlehem, Pa.
Read, Mrs. R. M....... ....175 Tremont Street, Boston, Mass.
Robinson, Mr. C. B.......N. Y. Botanical Garden, Bronx Park, N. Y. City.
Rondthaler, Miss E. W.............. Moravian Seminary, Bethlehem, Pa.·
Sanborn, Miss Sarah F...................54 Center Street, Concord, N. H.
Schumacher, Miss Rosalie........ ....Millington, N. J.
Seely, Mrs. J. A....................39 Caroline Street, Ogdensburg. N. Y.
Shreve, Mr. Forrest..... ....Johns Hopkins University, Baltimore, Md.
Smith, Mrs. Annie Morrill....... .. ..78 Orange Street, Brooklyn, N. Y.
Smith, Mrs. Charles C..............286 Marlborough Street, Boston, Mass.
Stevens, Mrs. M. L...................39 Columbia Street, Brookline. Mass.
Stevens, Mrs. O. H....................32 Pleasant Street, Marlboro, Mass·
Stockberger, Prof. W.W ....Bureau Plant Industry, Washington, D. C.
Streeter, Mrs. Milford B................113 Hooper Street, Brooklyn, N. Y.
Sweetser, Prof. Albert R ....University of Oregon, Eugene, Oregon.
Talbott, Mrs. Laura Osborne............"The Lenox," Washington, D. C.
Taylor, Mrs. A. P.......................................Thomasville, Ga.
Thompson, Miss Esther H..................... Box 407. Litchfield, Conn.
Thompson, Mrs. H. G......... .... Brooklyn, N. Y.
Towle, Miss Phebe M........... 19 Orchard Terrace, Burlington, Vermont.
Van der Eike, Mr. Paul................................Marine Mills, Wis.
Warner, Miss Edith A ... .78 Orange Street, Brooklyn, N. Y.
Wheeler, Miss Harriet..................... Chatham, Columbia Co., N. Y.
Wheeler, Miss Jane.......................248 Lark Street, Albany, N. Y.
Williams, Mrs. Mary E.....................1536 Pine Street, Philadelphia, Pa.
Williams. Mr. R. S.......N. Y. Botanical Garden, Bronx Park, N. Y. City.

VOLUME VIII. **NUMBER 2**

✖ MARCH, 1905 ✖

THE BRYOLOGIST

AN ILLUSTRATED BIMONTHLY DEVOTED TO

NORTH AMERICAN MOSSES

HEPATICS AND LICHENS

EDITORS:

ABEL JOEL GROUT and ANNIE MORRILL SMITH

CONTENTS

Entered at the Post Office at Brooklyn, N. Y., April 2, 1900, as second class or mail
matter, under Act of March 3, 1879.

Published by the Editors, 78 Orange St., Brooklyn, N. Y., U. S. A.

PRESS OF MC BRIDE & STERN, 97-99 CLIFF STREET. NEW YORK

THE BRYOLOGIST

BIMONTHLY JOURNAL DEVOTED TO THE STUDY OF NORTH AMERICAN MOSSES
HEPATICS AND LICHENS.

ALSO OFFICIAL ORGAN
OF THE SULLIVANT MOSS CHAPTER OF THE AGASSIZ ASSOCIATION.

Subscription Price, $1.00 a year. 20c. a copy. Four issues 1898, 35c. Four issues 1899, 35c.
Together, eight issues, 50c. Four issues 1900, 50c. Four issues 1901, 50c. Four Vols. $1.50
Six issues 1902, $1.00. Six issues 1903, $1.00. Six issues 1904, $1.00.

*Short articles and notes on mosses solicited from all students of the mosses. Address manu-
script to A. J. Grout, Boys' High School, Brooklyn, N. Y. Address all inquiries and subscriptions to Mrs. Annie Morrill Smith, 78 Orange Street, Brooklyn, N. Y. For advertising space address Mrs. Smith. Check, except N. Y. City, MUST contain 10 cents extra
for Clearing House charges.*

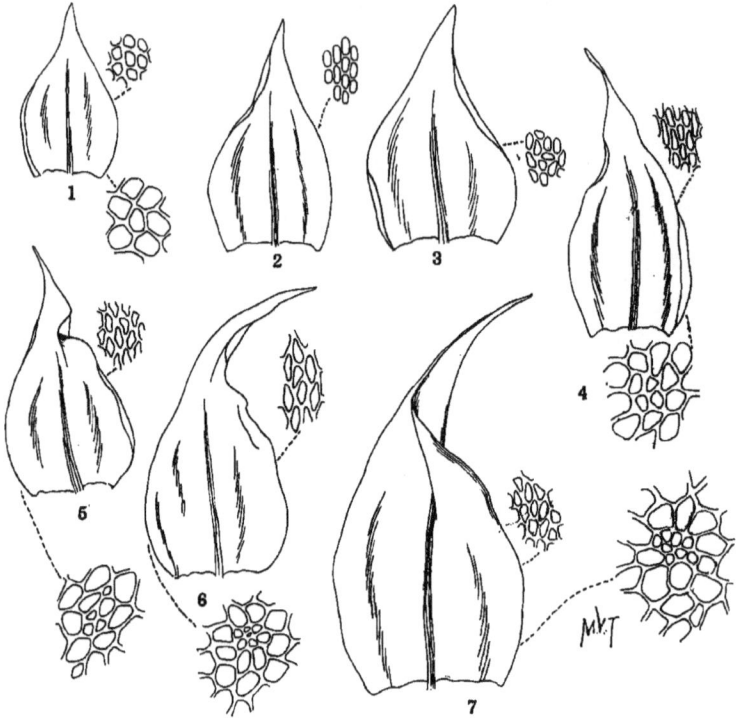

PLATE II.   Thuidium—Leaves, median cells and stem section.

THE BRYOLOGIST.

Vol. VIII.　　　　　　　MARCH, 1905.　　　　　　　No. 2.

A LESSON IN SYSTEMATIC BRYOLOGY.

Dr. George N. Best.

[Read at the meeting of the Sullivant Moss Chapter, Philadelpha, Pa., Dec. 31, 1904.]

While engaged in the study of some specimens of *Thuidium abietinum*, collected in Minnesota by Prof. J. M. Holzinger, I noticed that one of these differed from the ordinary forms of this species in that the leaves were somewhat larger, longer and more gradually acuminate, more strongly falcate-secund, and that the leaf cells were narrower and more elongated. Recognizing in these variations, at least in part, the characters on which Mitten had based his *Thuidium hystricosum*, and having an authentic specimen of this species at hand, a comparison was made, with the result of finding them nearly identical. The leaves, however, of *Thuidium hystricosum* were slighly larger and the leaf cells somewhat longer, but the differences in these respects were hardly appreciable. Not being fully satisfied that the Minnesota moss was indeed *Thuidium hystricosum*, I submitted it to Dr. Mitten, who considered it a form of *Thuidium abietinum*. Not comprehending how it was that two mosses, so nearly identical, should be referred to different species, I decided on an appeal, and in this instance to my herbarium, which contained about forty specimens of *Thuidium abietinum* from an extended range: that is to say, from Germany, Switzerland, France, Belgium, England, Labrador, various localities in the United States and in Canada, and as far north as the Yukon Territory. With few execeptions these specimens had been determined by well known bryologists. It may therefore be confidently assumed that they were correctly named. In fact, while *Thuidium abietinum* is but very rarely found in fruit, its specific characters are so well marked that no difficulty need be experienced in distinguishing it from all of the North American and European Thuidia, unless it be from *Thuidium hystricosum*, which I regard as simply a form of this species.

In my examination of these specimens attention was more especially directed to

1. The size and shape of the stem leaves taken from the middle third of the stem.
2. The size and shape of the leaf cells of these same leaves.
3. The presence or absence of a central strand in the stems.

As a tabulated statement of the results obtained from the examination of each of these specimens would be somewhat confusing, a few representative specimens have been selected by which it will be seen that we have here a series of forms, intergrading and ascending, and that the space between the extremes, that is between the lowest and the highest forms, is so covered by intermediate forms as to leave no doubt as to their being but variants of one specific type.

The January BRYOLOGIST was issued December 27th, 1904.

Fig. 1. Specimen from Crawford's Notch, New Hampshire; central strand none; leaves .8 mm. long, .5 mm. wide, ovate, gradually acute to abruptly acuminate, acumen straight or slighly curved; leaf cells somewhat uniform; median broadly oval to roundish oval, nearly isodiametric.

Fig. 2. Specimen from Northeastern Minnesota; central strand none; leaves differ from those of the preceding in being slightly larger, 1 mm. long, .6 mm. wide, more curved and longer acuminate; leaf cells larger and orbicular.

Fig. 3. Specimen from Labrador: central strand none; leaves broadly ovate, 1 mm. long, .8 mm. wide, gradually acute to abruptly acuminate, acumen curved; leaf cells somewhat irregular; median rhombic to oval-oblong.

Fig. 4. Specimen from Germany; central strand rudimentary; leaves 1.2 mm. long, .8 mm. wide: differ but little from those of the preceding; leaf cells somewhat longer and narrower, oval-rhombic to fusiform.

Fig. 5. Specimen from Niagara Falls; central strand rudimentary; leaves ovate-lanceolate, 1.4 mm. long. .8 mm. wide, slightly curved, narrowly acuminate; median leaf cells irregularly fusiform with subsinuate margins; apical cells linear-fusiform.

Fig. 6. Specimen from Northern Minnesota; central strand present but small and indistinct; leaves ovate-lanceolate, 1.6 mm. long, .9 mm. wide, narrowly acuminate, strongly curved: median leaf cells irregular, oval-rhombic to oblong-fusiform; apical linear-fusiform.

Fig. 7. Specimen from North Downs, England: central strand small, distinct; leaves ovate-lanceolate, 1.8 mm. long, .9 mm. wide, falcate-secund, long and narrowly acuminate; leaf cells not uniform, median oval-rhombic to fusiform; apical long linear-fusiform.—*Thuidium hystricosum* from Dr. Mitten.

To recapitulate: The leaves of the lowest form are ovate, .8 mm. long, .5 mm. wide, gradually acute to short acuminate, acumen straight or slightly curved, In the highest form the leaves are ovate-lanceolate, 1.8 mm. long, .9 mm. wide, falcate-secund, long and narrowly acuminate. The leaf cells in the lowest form are broadly oval, nearly isodiametric; in the highest form the leaf cells are irregular, the median oval-rhombic to fusiform and the apical linear-fusiform. In the lowest form the stems are without a central strand; in the highest form the central strand is small but distinct. In the intermediate forms the central strand is absent or rudimentary, and the median leaf cells pass from oval-rhombic to rhombic-fusiform or oblong-fusiform.

The parts of the plants which exhibit a fair degree of constancy or fixedness are the multiform paraphyllia, the spreading, pinnate, tapering branches, two-rowed on either side of the oval stem; the reddish, incrassate, porose basal cells of the stem leaves, the central papillæ on each surface of the leaf cells, longer on the lower than on the upper, and the midribs about three-fourths the length of the biplicate leaves.

The series here presented is one of the simplest of its kind and illustrates quite well the evolution of a specific type. Its individuals being usually sterile and its propagation effected almost wholly by vegetative means, the unknown factor of hybridism is thereby practically eliminated. While it would be useless to deny the possibility of still lower and still higher forms appearing, yet, so far as this series goes, it is without a break or a missing link in the evolutionary chain. Parts of series, broken series, the so-called groups, are always in evidence, but a series in which the connection of intergrading forms is so well kept up as in the one before us, is not so common. This difficulty however is probably to be attributed more to a scarcity of material, at least in many cases, than to an absence of such connecting forms in nature.

It is generally assumed that the lowest forms of a series are the oldest, and therefore properly represent the specific type, and that the highest forms are the latest and mark the evolution of this type. While this proposition is probably well founded, it scarcely admits of a demonstration: and while it is reasonably certain that plants appear in series, it is an open question whether these series are always ascending: possibly some are ascending, some comparatively stationary and others descending, in line for final extinction.

It is however more for its practical application in systematic work than for any theoretical consideration that this series is presented at this time, as I am fully convinced that if we are to make any considerable progress in systematic bryology, it must be along these lines. The younger bryologists among us know quite well what difficulties they have to contend with in determining the mosses they collect. If they succeed in tracing, with a reasonable degree of certainty, a given specimen to its genus, they are not infrequently baffled, notwithstanding the most persistent efforts, in making a satisfactory reference as to the species, two or three descriptions agreeing equally as well as any one. By not a little hard work and many bitter failures, the older ones among us have learned to discount these descriptions to their actual value, and are thereby enabled to approximate a determination with a reasonable degree of confidence.

In every system of classification, dealing with plants, the species is the unit of aggregation. It is therefore of the utmost importance that we should have a definite conception of what a species really is. In botany there is probably nothing so unscientific as the looseness with which species are usually made. Huxley defines a species as "the smallest group to which distinctive and invariable characters can be assigned." This is the traditional species, originating in a special act of creation, (whatever that may mean,) and continuing through all time with but limited variation, sexual trespass among the individuals of which being punished with annihilation. Huxley's species may be both logical and theological, but it is objectionable for the reason that it is too restrictive. Species with invariable characters are rare and probably do not exist outside of a single individual. Alphonse de Candolle says: "They are mistaken who repeat that the greater part of our species are clearly limited and that the doubtful species are in a feeble minority. This seemed to be true so long as a genus was imperfectly known and its species were founded on a few specimens, that is to say, were provisional. Just as we come to know them better intermediate forms flow in and doubts as to specific limits augment."

In this connection it may be well to quote Darwin in the chapter on "Variation under Nature" in *Origin of Species*. "From these remarks," he says, "it will be seen that I look at the term species as one arbitrarily given, for the sake of convenience, to a set of individuals closely resembling each other and that it does not essentially differ from the term variety which is given to less distinct and more fluctuating forms." It is needless to say that Darwin's pet theory was "natural selection." Before him other scientists had shown that in the organic world the indications pointed to an evolutionary process. Darwin claimed that the "survival of the fittest" was the keynote to this process. To better cope with the changing conditions continually presenting themselves in the world, species gave off varieties, these either disappeared in the course of time or became species, and these in turn gave off other varieties which likewise either disappeared or became species, and so on and on.

Under the guidance of such teachings it is no great wonder that the output of new species is as large as it is, for if Huxley's views be followed, calling for not only distinctive but invariable characters, and these not rarely the lowest in the scale of taxonomic values, the number of species must of necessity be largely augmented. On the other hand, if species is a term arbitrarily given for convenience sake to a set of individuals closely resembling each other, as Darwin would have us believe, the easiest way to dispose of a given specimen which does not readily fall in the line of recognized species, is to dub it a "new species" and let it conveniently pass at that. Thus it happens that mere scraps, without sexual organs or fruit, gathered from the "four corners," serve as the material out of which large batches of new species are made, apparently more for the glory of the makers than for the advancement of science. Taking advantage of Darwin's "convenience" for exploiting their ephemeral creations, they seem to lose sight of one of his requirements, in fact the principal one, namely, a "set of individuals." Possibly it is assumed that this "set" is found in a single specimen—I refer to mosses—but I am inclined to think that this assumption is a contravention of his real meaning. By "set" he probably intends to include a number of individuals not appearing in one tuft or from a single locality.

But Huxley's "group" and Darwin's "set" are alike objectionable in that they imply something artificial. In fact this is the dangerous reef upon which we are now stranded. Is it not about time for us to break away from these Linnean conceptions and to settle ourselves down to a more rational basis? Evolution does its work along well defined lines, not sporadically. If a species means anything, it means a series of individuals possessing certain distinctive but not invariable characters. In the higher orders of plants the sexual organs furnish excellent characters, not only for generic, but for specific distinctions. In the *Musci* the sexual organs are not rarely absent, and when present are not so distinctive. We must therefore rely on other parts for the differentiation of species. The leaves, and especially the leaf cells, are supposed to furnish valuable characters. We have seen, however, in the series we have examined that these were more or less variable, and that this variability depended on individual differences. The characters derived from the size and shape of the leaves and the size and shape of the leaf cells, are to be regarded as complementary to those more constant, and are not of themselves to be considered specific. In the species of some genera in which this variability is not so marked, the leaves and the leaf cells possess a higher degree of taxonomic value.

In his excellent work, *Die Laubmoose*, Limpricht places considerable stress upon certain anatomical characters derived from transverse sections of the stems and midribs. Whether these will prove more reliable than those more commonly employed or be more acceptable to bryologists in general is as yet an open question. Judging from my own observations I have no hesitation in saying that while undoubtedly valuable they vary more than would naturally be supposed, this becoming apparent when the extremes of growth are compared, and are therefore to be taken with due allowance. It cannot however be too strongly urged that in the delimitation of species the whole plant, and not a single part of it to the exclusion of the other parts, should be the subject of investigation.

From these considerations the fallacy of requiring each individual to conform in all its parts to the original type of a given species becomes evident. This type is just as likely to be one of the lower or one of the higher forms of the species in question, as it is to be an intermediate form. In either case exact duplicates should not be demanded. Mosses do not grow so much after a mathematical formula as some bryologists would have us believe.

It also appears that a single individual is not a species, although for descriptive and collective purposes it may be assumed so to be. When it becomes advisable, which is rarely the case, to make a new species of a single specimen, or when the material upon which it is based is poor and without fruit, there should be something to indicate these facts, as for instance a double dagger as a prefix to the name of the species. This would show that it was simply provisional and possibly not entitled to the rank given it. Just here it may be observed that new species are not so much needed, although they may be occasionally called for, as a better understanding of the limitations of those already in use. It may be further observed that to describe a species by comparing it with another by saying that the leaves in the one are a little longer or a little shorter, a little wider or a little narrower, the leaf cells a little larger or a little smaller, the pedicels a little longer or a little shorter, and this too without giving any measurements, is simply inexcusable. Species so described should not be recognized, neither should the makers of them.

The only way to acquire a true conception of a specific type is to study a number of specimens, the more the better, from as many different localities as possible. Extreme forms, intermediate forms, depauperate forms, all contribute to the series of which the species is the unit of aggregation. Subspecies, varieties, forms, may be necessary, but these will readily fall into their proper places when the specific type is once understood. The dimorphism which every species exhibits, and which is more apparent in some than in others, is only to be apprehended by a close study of the intermediate forms. From a taxonomic standpoint the recognition of these dimorphic tendencies is most important, as they are the marks by which the evolution of the type from its lower to its higher forms is shown.

In conclusion allow me to say that it is to be hoped that when this bloodless nomenclatorial war is over and when musty tomes and rotten types have done their worst and when personal aggrandizement has given way to the claims of science, more time will be found for the study of the making and the delimitation of species. Until that time it is to be feared that many a budding bryologist, full of life and hope, will be nipped by the frosts of many discouragements and driven to other fields in which more satisfactory results await less exacting labors.   Rosemont, New Jersey.

HOW TO COLLECT AND STUDY LICHENS.

BRUCE FINK.

Presented at the meeting of the Sullivant Moss Chapter, Philadelphia, Pa., Dec. 31, 1904.

Introductory.

It is a very real pleasure to the writer to be able to contribute to the meeting of the " Moss Chapter" something which he hopes may prove more or less interesting and suggestive. It was his privilege to be present at the meeting at Columbus, where the beginnings of the organization were made, and that meeting was so thoroughly enjoyable and instructive that he feels more keenly the loss at not being able to be at the present one. At the Columbus meeting, he expressed regrets that the lichenists could not have a similar society : but since that time both the bryologists and the mycologists have appeared to be so willing to give us room that we hardly feel the need of any separate organization. Especially safe is it to state that every American worker in lichenology feels grateful to the " Moss Chapter" for opening the pages of THE BRYOLOGIST for our articles on lichens. The work done there is already bearing fruit, and if the present writer can, by sending a paper to this meeting, aid those who have shown an interest in us and our work on the lichens, he will at the same time serve his own ends and those of lichenologists generally quite effectually. So to the matter of collecting and studying lichens without further introductory statement, except to say that only a popular statement can be given in the short time.

Collecting.

Lichens can be collected at any time in the year, but many of them are more likely to show the spore characters better when collected in the fall. They may be collected also on any kind of a day, but more effective work will be done on pleasant days, while many of the minute forms are more easily detected when damp and therefore brighter. The beginner will find the more conspicuous foliose and fruticose species first, and will be very likely to begin with the lichens of the trees. After finding a few of the most conspicuous forms, he will think that he is nearly done, and yet every time he goes over the same ground something new will appear. This is the uniform experience, and even the most keen-eyed lichenist finds much of interest after he has been over a limited area several times. The difference between him and the beginner is that he knows from experience that he does not detect all the first few times over a spot; while the beginner has to learn this, little realizing how few of the many species he sees at first and how poorly he distinguishes differences in lichens, at first thinking that three or four forms are all one, when perhaps they do not even belong to the same genus. But careful study will soon begin to improve the powers of observation, and the work will grow and the interest increase day by day. In continuing to work on a small area till it has been looked over a dozen times or more, one should attempt to find every substratum that might bear lichens and take into account all the varied conditions of light, shade, and moisture, which cause so much of the variation in species. Look carefully on old dying trees, trees in good condition but old with rough bark, and younger trees with smooth bark, for rocks in shade and rocks exposed to sunlight, the outcrops and the boulders and for shaded and exposed earth. Then look for any species of lichens that seem to prefer a particular genus or species of tree or a particular kind of rock or earth. And when all this is done in an average region, the beginner should be able to find from 100 to 150 lichen species and varieties within five miles of his home, while in some localities such work carried on for two or three years should give the student more than 200 lichens.

Collecting Outfit.

But before the first trip is made the student will want to know what to carry with him on a collecting trip. A good knife is needed to take the lichens from the trees with as little of the bark as possible, and a geologist's hammer and a good cold chisel, especially tempered for the rocks to be chipped, to get the rock lichens with as little of the rock as possible. Then a hand lens is needed to enable one to detect differences in lichens in the field so as to know as far as possible whether he is duplicating too much. With the lens one soon comes to detect differences in both surfaces of the larger thalli, the nature of the exciple and disk, the upper surface of crustose lichens and many apparently slight microscopic differences in minute lichens, which might otherwise be thought to be the same in the field when they do not even belong to the same genus. A bag, basket or vasculum must be carried to contain the specimens, and a lot of paper or envelopes so that each kind of lichen may be wrapped separately as it is collected. Then there must be a pen or pencil so that careful notes may be placed on each envelope or in each packet, showing the date of collecting, the name of the substatum and the surrounding conditions as to light, moisture and shade, or any other data that may be desirable in a particular instance. A sponge is sometimes very handy for moistening certain lichens in dry weather so that they may be easily separated from the substratum, and a small bottle of water may easily be carried to wet the sponge from time to time. Delicate specimens from rocks or earth should be wrapped separately in any old paper to prevent breaking or abrasion.

Where to Collect.

Go to some well wooded area if such is at hand and begin work as already suggested on the large foliose lichens of the trees. Do not go on a long tramp, but as soon as you are in the woods, collect the lichens that are growing all around. It is too common an error to start on a long tramp, and many a collector walks for miles and goes by the lichens on every side because he thinks he will find something better just ahead. The result is a long walk and few specimens. If woods are not at hand and rocks are, they will serve for a beginning but beginners always make bad work of chipping rocks. Do not carry home a cord of rocks: but get pieces just large enough so as to get the lichen wanted complete, or at least enough of it to show the border of the thallas on one side. In many areas, especially in the pineries, one may well begin with the earth lichens. They are there easy to find and collect, but often require a good deal of care at home. If so unfortunate as not to live near woods or rocks, lichens may still be found on old fences and on trees planted along roads and in yards, etc. In the woods, be sure to examine old logs and stumps, corticate and decorticate, sound and rotten, standing, erect and prostrate. In examining boulders and pebbles, look at all sizes; and on the larger ones expect different species near the ground from those growing at the top where there is less of moisture. As stated in a preceding paragraph, look out for different kinds of trees, rocks and earth. Uninhabited and undisturbed wooded regions are the best places in the world for lichens, but there is no place where they may not be found, for they occur even on the prairies and about the large cities.

Aids at Home.

We will suppose that the first collection is made. It matters not what the species are for the first time, but they are very probably the most common of lichens, just as they should be. Perhaps the specimens are small and fragmentary, but the collector will soon learn by experience that it pays to get good material, and if he is to exchange later on or send away for determination, to get it in abundance. No warning will have much weight till he has run out of some rare material in exchange, or has frequently been told by one of more experience that his material sent is too fragmentary for determination. But leaving this for the present, what is needed at home in order to work effectively? If possible, there should be a table permanently placed for work. On it should be a microscope, magnifying at least 550 diameters, a good sharp razor for cutting sections of fruit and thalli, pith in which to cut the sections, a small bottle of water and another of potassium hydrate, slides and cover glasses, an eye-piece micrometer for measuring the size of spores, and some volume which contains descriptions of all the common lichens. The sections are to be cut in the elder pith with a very sharp razor, and they must be thin enough so as to be more or less transparent under the microscope. These sections may be mounted directly in water, and in most instances no other solution is needed. However, if the sections are not clear, the water may be drawn out gradually and replaced by the potassium hydrate, placing a drop of this solution at one side of the cover glass and a bit of absorbent paper at the other side to take up the water. Filter paper serves this purpose best. If the asci, spores and paraphyses do not come out distinctly with this treatment, the section may be carefully crushed after the character and color of the exciple or exciples, the hypothecium and the hymenium are all studied. Then if still unsuccessful, some stain may be applied. Iodine solution will serve to differentiate between the asci and the paraphyses as it stains them differently and often brings out the branching of the paraphyses beautifully. Some experince will enable the beginner to get just the best strength of iodine solution, but one grain of iodine, three grains of iodine of potassium and one of pure water makes a very good combination. The sections are almost sure to be too thick at first, but experience will remedy this difficulty. The razor should be sharp enough to cut a section of the pith thin enough so that it will float in the air, and then the section of the lichen or lichen apothecium will be so thin that one will often need to place his slide on white paper in order to see the sections, which are to be transferred from the razor to the slide by means of a small camel's hair or other similar brush. To insert the material to be sectioned into the pith, cut a slit through a radius of the pith from one end down an inch or more. Then taking a portion of thallus or fruit 2 or 3 mm. across, insert into the opening in such away as to be able to cut in the direction desired. If the fruit is larger than 3 mm. in diameter, it is still best not to try to cut larger sections, but an edge of the apothecium is to be included in the section, and it is permissible to section one whole, cutting through a diameter so as to see the structure. However, this section is likely to be of little use for any careful work. Do not attempt to cut all the way across the pith at every cut, but rather to get very thin sections of small portions of the upper surface of the pith, including a section of part or all of the lichen structure to be studied. This procedure will soon render the upper surface of the pith uneven, when a complete section may be taken to level it. Many beginners will not think all of this advice necessary; but all will appreciate it after a few trials, and will wish it were possible to make matters much plainer than can possibly be done in any written directions. As to the razor, it must be of good quality, not too thick, and is better if hollow ground on one side only. Then keep it sharp, sharp! sharp!! Do not sharpen a moment and then resume work with a contented air, but see if it will readily cut sections of pith that are scarcely visible when floating in the air. If not, it is not in condition for cutting sections of lichens. Then there should be always at hand on the table a metric rule, for the larger measurements of thalli and fruits, which of course can not be made with the micrometer.

There are many other things that might be stated, but too much is confusing to the beginner. At first, throw away unfruited specimens, unless you have material with which to compare, but later, after you know something of lichen species and think you have a sterile one different from any of the fertile ones, determine it or sent to an expert. If you have no microscope, you can still do some good work with fifty or one hundred authentic specimens with which to compare and the series of articles published in the BRYOLOGIST by Mrs. Harris. The beginner will not need foreign literature, but the books written by Tuckerman, Schneider and Willey will be found helpful if they can be obtained. As you gain in experience, difficulties will clear up, and you will find ways of your own and perhaps better ones than some of those suggested above, for no two people work just alike. At least do not collect without attempting to determine as best you can, Send specimens to an expert with your determinations stated on every envelope, even if you are no farther than the genus or even the family. You must do this for the sake of the satisfaction of it and the strength that it will give.

The Study at Home.

But given this table and apparatus and some directions regarding use. Just what shall be studied? Regarding the thallus there must be careful observation of form, size, color, method of attachment and general relation to the substratum, nature of the surface as to whether smooth, wrinkled, chinky, areolate, verrucose, etc., the margin as to whether entire, wavy, or lobed, etc., and finally the cross-section must be made and carefully studied. Then turning attention to the fruit, the general form and size must be carefully noted, the form and color of the disk, the nature and duration of the exciple or exciples, and the manner of attachment of the fruit to the thallus. Then the sections may be resorted to in order to ascertain the nature of the exciple, the hypothecium, the paraphyses, the asci, and the spores. And, finally, it will be found to be an excellent exercise to attempt to write a description of a lichen occasionally bringing out all the points observed. After a few descriptions have been written, those in manuals of lichens will mean more to the student, for they will not appear so vague as soon as the powers of observation and discrimination are thoroughly developed. Sometimes one can put some special "ear mark" into a description, but often two species are somewhat different in a number of points, but not very much so in any one particular respect. In such instances, the attempt to show the special "ear mark" will be a failure, and the decision between the two species may be by no means easy for the most competent student of lichens. The beginner must always see the spores in the asci, as he is otherwise very likely to get the spores of some other lichen occasionally and make a stupid failure in the determination. And the student must be warned not be expect to find sections like many of the drawings in some manuals of lichenology. Many of these figures are diagrams, which show what might be seen in ideal sections. They serve their purpose, but the student will usually have to be content with seeing things much less distinctly. Finally, after the beginner has done his best, he will often have to be satisfied with tracing his plant to the genus or family rather than to the species; but he need not be discouraged at this, for experience will make him more and more able to determine species. Every manual of lichens has some peculiarities that need explanation. And perhaps the uses of the terms *pale* and *cloudy*, as applied to the hymenium and the hypothecium in the descriptions issued by the present writer, need special explanation. In ordinary sections the lightest colored areas in these tissues seem whitish, whereas, if the sections were thinner, they might appear perfectly hyaline. These areas have been called pale. Then in some other lichens similar sections are somewhat denser in these areas so that in sections of ordinary thinkness there is a darkish cast that really appears like the color of clouds. Then plainly enough, a section will tend toward a pale appearance if thin, and is more likely to be cloudy if thick. And in the interpretations of these and all other colors seen in sections, some allowance must be made for the thickness of the section. Of all the diagnostic characters given in descriptions, perhaps those regarding the paraphyses have least value. The common statement is about thus, paraphyses simple or rarely branched, commonly enlarged and brownish toward the apex. This answers for the great majority of lichens with little modification but when the statement varies considerably from this form, the paraphyses are of more consequence in determination. Also, it should be said that in measurements, macroscopic and microscopic, there is usually no special effort made to reach the rarest extremes in sizes. So the student need not be surprised at finding occasionally larger or smaller measurements than those given. However, the extreme sizes must not vary greatly from those of the manual used.

The Herbarium.

Specimens once determined should be carefully dried so as to avoid moulding, and the larger ones are to be pressed in the same way that higher plants are pressed, placing the specimens in the press, not soaked with water, but just damp enough to press well. The crustose and closely adnate foliose species seldom need pressing. Earth containing small foliose or crustose species must be saturated with mucilage, which will keep it from crumbling in the herbarium and destroying the specimens. All but the rock specimens keep well in the ordinary herbarium envelopes, and even they are often kept in the envelopes also. But if one is not very careful not to get large pieces of rock, it is usually necessary to resort to stronger and larger envelopes or pasteboard boxes for these rock lichens. Delicate specimens as *Caliciums*, and members of some other genera had better be glued to the bottom of small boxes in such a manner that the delicate lichens will be out of contact with anything else than the substratum on which they grew. The envelopes may be mounted on ordinary herbarium paper, but brown paper is very commonly used by the lichenists of Europe, both for envelopes and mounting paper. This paper does not show dirt as does the white paper. Many paste all specimens to paper if removed from the substratum; but if this is done, part of the material must be placed ventral side upward so that both sides of the thallus may be seen. This method helps to prevent breaking the brittle thalli, but interferes somewhat with the study of the specimens. All specimens in the herbarium must contain careful data such as those suggested to be taken in the field. Finally, it is not possible that a short paper should contain all the suggestions that are valuable in the collecting and study of lichens, but it is hoped that those given may enable the members of the chapter and others to work somewhat intelligently, while gaining that experience which is more valuable than any directions that can be given.   Grinnell, Iowa.

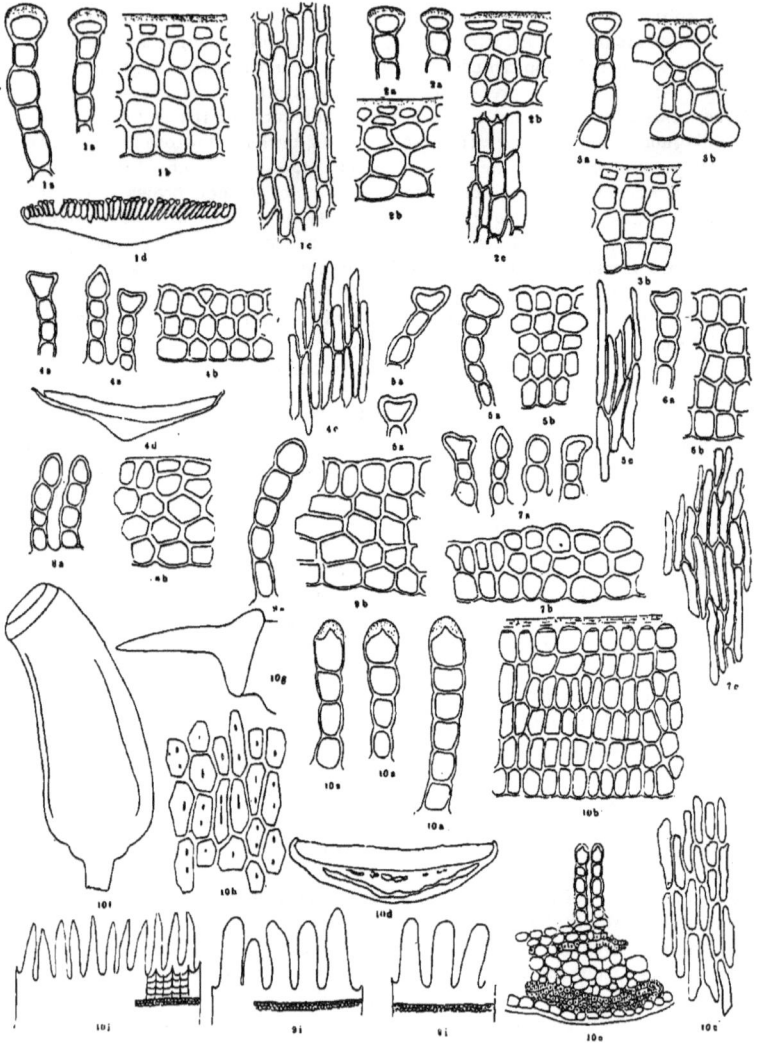

PLATE III. Polytricha.

SOME RECENTLY DESCRIBED NORTH AMERICAN POLYTRICHA.

JOHN M. HOLZINGER.

In Limpricht's Laubmoose, Band II, p. 853, the author has this note:
" P. 618: No. 618. Polytrichum Ohioense Ren. et Card. in Rev. Bryol.
1885, p. 11 and 12, and in Coult. Bot. Gaz. XIII. p. 199, t. 17 (1888) has the
priority over Polytrichum decipiens!"

P. decipiens is described by Limpricht in 68. Jahresb. d. Schles. Ges.
f. vaterl. Cultur 1890, II. p. 93; also in Laubmoose II, 1894, p. 618. In 1895,
in August or September, appeared the note on p. 853 of Laubmoose, above
cited. This shows that at the time of that writing Limpricht considered his
P. decipiens identical with *P. Ohioense*, reducing his name to a synonym of
the latter. Then in 1900, in Bot. Centralbl. XXI. Jahrg., No. 50, Prof.
Harald Lindberg showed that the two plants are really separate, This view
of the situation is honored by Limpricht in Laubmoose, Bd. III, (1903) p.
800, where he again separates his *P. decipiens* from *P. Ohioense.* While *P.
Ohioense* appears to be exclusively North American, *P. decipiens* is found
both in Europe and this country. Prof. Lindberg cites the following localities for *P. Ohioense:* Wisconsin, Milwaukee, leg. Lapham (Ex. Herb. Cardot). Lake Michigan, leg. Lapham (Ex, Herb. Cardot).

Illinois, Chicago, 1888, J. Röll (No. 1811). Edgewater near Chicago,
20. 9, 1888, J. Röll (No. 1815).

New Jersey, Hoboken, 8, 1898, P. T. Cleve.

Massachusetts, Milton, Blue Hill, 2, 6, 1898, 28, 8. 1898, 26. 12. 1898. G
G. Kennedy.

District of Columbia, Rock Creek, 10. 6. 1894, J. M. Holzinger.

Renauld and Cardot in their check list of Musci Americæ Septentrionalis,
p. 41, give the range of *P. Ohioense* as: Canada, Northern, Eastern and
Central States, questionably British Columbia.

The range of *P. decipiens* in North America is given by Lindberg as follows : "Prince Edwards Island (as *P. Ohioense* Ren. et Card. in Can.
Musci, No. 221). To this species belongs also No. 323 in Sullivant et
Lesquereux, Musci Bor. Americani, named *P. formosum* Hedw. Mr. Cardot refers this form in Botanical Gazette. Aug. 1888, to *P. Ohioense.* The
specimens in Musci. Bor. Am. are without locality." To this may be added
a station in Minnesota, Miss Sarah O'Meara collecting near St. Charles,
Winona Co., a plant the writer has referred to *P. decipiens.*

The differences between these two species as pointed out by Lindberg in
Bot. Centralbl. XXI. Jahrg., No. 50, are as follows:

" P. OHIOENSE Ren. et Card.

"The lamellæ of the leaves when seen from the side have a plane margin not crenulate, strongly thickened, more or less distinctly papillose with
marginal cells much smaller than the others : the marginal cells in cross-
section always concave, very much alike, strongly thickened especially on
the outer wall. On the back of the leaf the cells are for the most part
arranged lengthwise (i. e. in longitudinal rows), the cells of the sheath of the
leaves shorter and broader."

" P. DECIPIENS Limpr.

" The lamellæ of the leaves when seen from the side have a crenulate margin, not or only slightly thickened, and not papillose, and all the cells are nearly equally large ; the marginal cells of the lamellæ in cross section are usually unlike each other, but for the most part slightly emarginate. On the back of the leaf the cells are for the most part transversely arranged. The cells of the sheath of the leaves are longer and narrower."

In the same paper Prof. Lindberg describes

" POLYTRICHUM ANGUSTIDENS n. sp.

" Plant 4 cm. high, brown green, stout, simple or usually branched, densely leafed, not radiculose. Leaves when dry twisted, recurved or erect-open, when moist lower leaves erect-open, upper leaves recurved slightly ; the lamina 10 mm. long, at the base about .7 mm. wide, gradually narrowed into a short brown sharp and denticulate point : leaf base sheathing, 1.8 mm. wide, somewhat shining when dry. Leaves in cross-section obtusely keeled, the costa on the back somewhat prominent, occupying nearly the entire lamina, with a thick dorsal bundle of stereid cells across the leaf, the ventral bundle less developed and interrupted, the dorsal cells rather large with their outer walls thickened. Lamellæ about 46, .07 to .1 mm. high, closely crowded, erect, built up of one layer of cells (4-6 deep), greatly thickened on the margin, which is plane, not crenulate, but lengthwise striolate, the marginal cells in cross section not or little larger than the rest, otherwise similar to them ; convex above, papillose, the greatly thickened walls crescent-shaped. Seta straight. stiff, 56 mm. thick, purple, about 50 mm. long. Capsule oblique, microstome much larger at base. 5.7 mm. long, at base 2.3 mm. in diameter, at the mouth only 1.5 mm., sharply quadrangular, with a very distinct hypophysis having stomata, with cells of exothecium haxagonal and grooved. Operculum conic, 2.2 mm. long, obliquely long-rostrate. Basilar membrane of peristome .1 mm. high, teeth 64, narrow, sharp pointed, pale, papillose, .2 mm. high and about .035 mm. wide. Spores green, very smooth, pellucid, 8.8-11μ in diameter.

" This is a very fine new species very well marked by the characters above set forth."

Type station near Hope. Kootanai Co., Idaho, collected by Dr. J. H. Sandberg in August, 1892. (No. 1121, in Contributions U. S. National Herb., Vol. III, No. 4, p. 272, as *P. attenuatum* Menz.) Type in the Nat. Herb., Washington, D. C., as No. 1121 of Dr. Sandberg's collection. A duplicate of this plant has been sent by the writer to Dr. V. F. Brotherus as *P. formosum* Hedw. (old No. 137, new No. 1121). And this is the plant described above.

It seems best to publish with this note Prof. Lindberg's plate elucidating the differences between the several species of Polytrichum referred to.

Fig. 1. *Polytrichum Ohioense* R. & C., Lake Michigan, leg. Lapham (Herb. Cardot).

Fig. 2. *P. Ohioense* R & C., New Jersey, Hoboken, leg. P. T. Cleve.

Fig. 3. *P. Ohioense* R. & C., Dist. of Columbia, Rock Creek, leg. J. M. Holzinger.

Fig. 4. *P. decipiens* Limpr., Bohemia, Böhmerwald, leg. E. Bauer.

Fig. 5. *P. decipiens* Limpr., Sull. & Lesq., Musci Bor. Amer. No. 323 (as *P. formosum*).

Fig. 6. *P. decipiens* Limpr., Finlandia, Isthmus Kerelicus, par Metsapirtti, leg. Harald Lindberg.

Fig. 7. *P. decipiens* Limpr., Finlandia, Isthmus Kerelicus, par Sakkola, leg. Harald Lindberg.

Fig. 8. *P. attenuatum* Menz., Finlandia, par Lojo, leg. Harald Lindberg.

Fig. 9. *P. gracile* Dicks., Finlandia, Helsingfors, leg. S. O. Lindberg.

Fig. 10. *P. angustidens* Lindb. fil n. sp. leg. J. H. Sandberg (U. S. Nat. Herb).

a. lamella in cross-section X 280. b. lamella, side view X 280. c. cells from the middle part of the sheathing leafbase X 130. d. cross-section of eaf X 40. e. cross-section of leaf X 130. f. capsule X 7, g. operculum X 9. h. cells of exothecium X 180. i. part of peristome X 40.

Winona, Minn.

NOTES ON A COLONY OF HEPATICS FOUND ASSOCIATED ON A DEAD FUNGUS.

CAROLINE COVENTRY HAYNES.

(The fungus was exhibited, with original drawings to illustrate, at meeting of the Sullivant Moss Chapter, Philadelphia, Pa., Dec. 31, 1904.)

The fungus *Fomes fomentarius* blackened and sodden, was found while collecting in the Adirondacks on the Adirondack League Club Tract, attached to a decayed log, once a yellow birch, lying in a bog ; it caught and held the moisture and the hepatics and mosses found growing upon it testified to its desirability as a residence, from their standpoint, quite as they would have done had it been their usual habitat. It was an interesting task to examine it carefully, mounting specimens and determining the ten species found growing upon it, the majority of them moisture loving.

There were a few plants of *Scapania curta* (Mart.) Dumort., a quite rare species, the finding of which is always a delight. Of the three Cephalozias: *C. curvifolia* (Dicks.) Dumort. is easily recognizable with a hand-lens, tiny as it is, the clearest three-angled perianths, and the saccate leaves, with clawlike lobes, are quite unlike anything else: *C. lunulæfolia* Dumort. equals *C. media* Lindb., while resembling *C. connivens* (Dicks.) Lindb., has smaller leaf cells and the perianth mouth is short ciliate, that of *C. connivens*, long ciliate: *C. serriflora* Lindb., has usually been known in this country under the names *C. catenulata* Spruce and *C. Virginiana* Spruce. *C. Virginiana* is now regarded by Dr. Evans as being "scarcely distinct" from *C. catenulata*. In his "Notes on New England Hepaticæ" Rhodora, Vol. 6. 1904. p. 173, Dr. Evans makes clear several facts regarding this species that Lindberg called *C. serriflora*, and he, like Dr. Evans, follows Jack and others in the use of it. Spruce in 1882 named and described *C. catenulata*, mentioning denticulate perichætial bracts, which however, in rare cases were nearly or quite entire. Recent European writers believe that Spruce in reality described two distinct specific types; those with entire bracts being the true *Jungermannia catenulata* of Hübener which is unknown in North America, those with denticulate bracts being *Jungermannia reclusa* of Taylor. Now Taylor, in his original description in 1846, does not mention the bracts, and the plants thus labeled in his herbarium include several distinct species in poor condition : he probably did not thoroughly understand the species, therefore, the use of Lindberg's name, *C. serriflora*, is to be commended. I quote again from " Notes, etc." : " From other species growing on rotten logs it may be distinguished by its widely spreading and deeply bifid leaves, the acute divisions being straight or slightly connivant; by its leaf cells with uniformly thickened walls; by its dentate or denticulate perichætial bracts, and by its thin-walled, three-angled perianth with ciliate mouth." Range from Canada to the Gulf States.

In the damp depressions of the fungus I found *Riccardia latifrons* Lindb. closely crowded together; with narrow thallus, palmately divided. *Jamesoniella autumnalis* (DC.) Steph. the *Jungermannia Schraderi* Martius of Grey's Manual sprawled here and there, sterile specimens look a little like *Odontoschisma prostratum*, as Dr. Evans once pointed out to me in a letter, only "its leaves are less distinctly margined and show larger cells." *Jamesoniella autumnalis*, *Kantia trichomanis* (L.) S. F. Gray, and *Lophozia ventricosa* (Dicks.) Dumort. were the largest members of the colony, individual specimens being easily recognizable with a hand lens. There are excruciating nomenclatorial changes being endured by the Kantias in Europe but the contagion has not yet spread over here. They can be found in May in the Adirondacks bearing capsules which open in four curiously-twisted flame-colored lobes at the end of a very long and slender seta.

Dr. Howe gives a good key to the Lophozias in his "Hepaticæ and Anthocerotes of California," Memoirs of the Torrey Botanical Club, Vol. 7. p. 104, 1899. I will insert the description of *L. ventricosa:* "Leaves two (rarely three) lobed 1/5 to 2/5 their length, close or approximate, the lobes acute, occasionally apiculate, rarely subobtuse, entire, the sinus broad; median leaf cells 24–32μ." His account of the species is most interesting.

Lophozia incisa (Schrad.) Dumort. is one of my favorites, the finely-drawn fluted spinulose-dentate leaves clasping the stem remind me of stiff lace ruffs in the reign of Queen Bess. And of the filamentous *Blepharostoma trichopyyllum* (L.) Dumort., how shall I speak! The hand-lens shows the spider-web-like leaves and stems flung gauzily here and there over the larger species, but it requires the high power of a microscope to discover its structure. Besides these ten hepatics, there were two mosses, immature plants of *Hypnum reptile* and *Dicranum* sp?

New York City.

RHACOMITRIUM HETEROSTICHUM GRACILESCENS.

I have found on the top of Bald Mt. (altitude about 1.100 feet), in Camden, Maine, a sterile moss that is believed by Mr. E. B. Chamberlain and Dr. G. N. Best to be *Rhacomitrium heterostichum gracilescens* Br. & Sch., although there are slight differences between this and the type. This moss has been reported from Canada under other names, collected by Prof. John Macoun, but not before from Maine, and probably not from the United States. The sides of Bald Mt. are wooded, but the top is an immense bare ledge. In a depression of this ledge the moss was growing. I have enough of the moss to supply the usual request for offerings. If more should be called for I will furnish as soon as a fresh supply can be had.

ALICE L. CROCKETT.

SULLIVANT MOSS CHAPTER NOTES

The following ten names have been added to the list of Chapter Members since January 1st, making the total number 143.

M. Henri de Poli, 45 Rue des Acacias, Paris, France; Mr. Wm. Edward Nicholson, Lewes, Sussex, England: Miss Julia P. Brigham, 138 Pleasant St., Marborough, Mass.; M. Georges Lachenaud, Nexon, Haute-Vienne, France; Köno Gakuichi, Hiroschima, Japan: Mitsujiro Kawasaki, Ise, Japan; Genji Koyama, Kioto, Japan: Mr. George P. Annand, 39 Brown St. Waltham, Mass.; Mr. A. S. Foster, Hamilton, Wash.: Prof. T. J. Fitzpatrick, Iowa City, Iowa.

REPORT OF THE PHILADELPHIA MEETING.

The third meeting of the Sullivant Moss Chapter was held on December 31, 1904, at 1.30 p. m., in the Academy of Natural Sciences, Philadelphia. In the absence of the President and Vice-President the meeting was called to order by Dr. A. J. Grout, and Dr. A. W. Evans was elected Chairman. The reports of the retiring President, Prof. John M. Holzinger, and Secretary, were read by the latter, and the presentation of papers followed. The first paper to be read was a very interesting one by Dr. George N. Best on "A Lesson in Systematic Bryology," illustrated with slides and drawings. Dr. Evans followed with a paper on "Leafy Hepatics," with drawings illustrative of their generic differences. Dr. A. J. Grout's talk on "The Use of the Hand-Lens in Studying Mosses and Hepatics," with a practical demonstration with slides and hand-lenses, proved that much excellent bryological work can be accomplished without the aid of a compound microscope. Mrs. Elizabeth G. Britton gave an entertaining talk on "Nassau and Florida Mosses," illustrated with drawings. As the time was limited Dr. Evans read only the introduction of the paper sent by Prof. Bruce Fink, "How to Collect and Study the Lichens." The members then voted that a greeting be sent to the Vice-President, Mrs. Harris, whose recent serious illness prevented her attendance. The meeting was then adjourned and the rest of the afternoon was spent in examining the Chapter Exhibit. This was open to the public from December 28th to the 31st.

On the evening of December 29th the Academy of Natural Sciences tendered a reception to the members of the American Association for the Advancement of Science and to the visiting botanists, and there was an informal gathering of the members of the Chapter and their friends at that time. The walls were hung with mounted specimens of mosses, hepatics and lichens from the Chapter Herbaria and private collections. A series of type specimens of hepaticæ, with slides for the microscope, complete files of THE BRYOLOGIST, artists proofs, books, pamphlets, photographs of botanists, "Lists" by members of the Chapter, and some beautiful photographs or "portraits" of lichens and lichenists by Mr. G. K. Merrill, added much to the interest of the Exhibit. By the courtesy of the officers of the Academy some of their rare collections of mosses and hepatics were also on exhibition. Of special interest were the Mühlenberg and Drummond exsiccatæ.

Though the attendance was not as large at the Saturday afternoon meeting as it doubtless would have been had it been possible to arrange for an earlier day, still it was most interesting and many who were absent expressed their regret at being unable to attend. The reports of work accomplished during 1904 showed an ever increasing interest in the Chapter. Since the yearly report was written it has been decided to admit foreigners to membership, and we now have representatives from Japan, France and England on our lists of members.   Respectfully submitted.

MARY F. MILLER,
Secretary.

MY VALEDICTORY.

The time is fast approaching and now is, when I shall no longer have the honor nor perform the tasks of the first officer of the Sullivant Moss Chapter. The retrospect to this honor is a real pleasure—quite as much as the revelry in the midst of presidential duties. By what seems to have been a special dispensation there have not been quite so many calls for help from young members during my second term of office as there were during my first term. Still there has been no chance to feel lonesome, and one or two lots of mosses are even yet not attended to: these will be disposed of before the end of the winter. I have enjoyed to be called upon for help and shall be slow to refuse assistance. Necessarily, in the midst of busy school duties this work on mosses frequently suffers protracted delay. And I thank all my correspondents for their uniform patience.

In order to lighten the work of my successors may I not make some suggestions additional to those made a year ago? I then asked that correspondents refrain from sending mere bits or scraps. I consider that every one asking help should consider it his duty and privilege to send a fairly sufficient quantity of a species of moss to be determined, first, in a respectable properly folded pocket of proper size; second, to have each pocket labelled in a uniform way with the sender's and collector's name, with locality and with date. If the sender does not do this with each pocket then I must do it for him, and this I have always felt to be an uncalled for imposition, due it may be to the sender's thoughtlessness. And if, after this very blunt reminder, if seekers for help do not comply with so eminently fair a requirement, I consider that my honorable successors will be fully justified, unless there are special reasons for leniency, to consign all mere scraps, simply numbered, and carelessly wrapped in a piece of dirty grocery paper, unceremoniously to the waste basket. Only good material put up in a business-like way and sufficiently labelled should receive attention.

It is with sincere regret that I record my inability to meet with the members of this very enjoyable Moss Chapter on December 31, 1904, in Philadelphia. Both the distance and conflict with our State Teachers' meeting forbid that I attend. How happy I would be to really shake hands once more with my numerous very friendly correspondents, such as Dr. Best, Dr. Grout, Mrs. Smith, Mrs. Lowe, Miss Miller, and others. As it is, I have to be content with sending a mere friendly greeting to all present, through our Secretary, Miss Miller. To *all* I also send a Happy New Year!

JOHN M. HOLZINGER, Retiring President,

Winona, Minn.

OFFERINGS.

(To Chapter Members only. For postage.)

Mr. Edward B. Chamberlain, 1830 Jefferson Place, Washington, D. C. *Aphanorhegma serratum* Sulliv., c.fr. Collected in Maryland.

Miss Alice L. Crockett, Camden, Maine. *Rhacomitrium heterostichum gracilescens* B. & S., st. Collected in Camden, Maine.

Mrs. J. D. Lowe, Noroton, Fairfield Co., Conn. *Hypnum fertile* Sendt., c.fr.; *H. uncinatum* Hedw.. c. fr.; *Aulacomnium androgynum* Schwaegr., c.fr. Collected in Maine. *Thuidium delicatulum* (L.) Mitt., c fr.; *Pogonatum tenue* (Menz.) E. G. Britton, c.fr.; *Bæomyces roseus* Pers. Collected in Connecticut.

Mr. H. Dupret, Seminary of Philosophy, Montreal, Canada. *Bryum roseum* Schreb., c.fr; *Bartramia pomiformis* Hed., c.fr. Collected near Montreal. U. S. Postage accepted.

Miss Mary F. Miller, 1109 M Street, N. W., Washington, D. C. *Lepidozia sylvatica* Evans. Collected by A. J. Grout in Flushing, L. I. *Timmia megapolitana* Hedw., c.fr.; *Didymodon rubellus* (Hoffm.) B. & S., c.fr. Collected in Vermont.

Miss Caroline C, Haynes, 16 East 36th street, New York City. *Cephalozia curvifolia* (Dicks.) Dumort.; *Jamesoniella autumnalis* (DC.) Steph. Collected in the Adirondack Mts., N. Y.

Mr. G. K. Merrill, 564 Main street, Rockland, Maine. *Cetraria cucullata* (Ball.) Ach. Collected on Mt. Washington, N. H.

Mrs. Annie Morrill Smith, 78 Orange street, Brooklyn, N. Y. *Peltigera venosa* (L.) Hoffm., c.fr. Collected by Mr. A. J. Hill, British Columbia.

A CORRECTION. Mr. Rapp writes that the moss offered by him in January, 1905, should have been given as *Funaria hygrometrica* var. *patula*.

NEW OR UNRECORDED MOSSES OF NORTH AMERICA.

By J. Cardot and I. Thériot.

Translated and condensed from The Botanical Gazette, May, 1904.

DESCRIPTIONS OF NEW SPECIES GIVEN IN FULL.

BARBULA BAKERI Card. & Ther.

Dioicous, loosely caespitose, olivaceous or lurid green, 3-8 mm. high. Leaves erect and contorted or crispate when dry, patulous-squarrose when moist, 1.4-1.5 mm. by 0.6-0.7 mm., rather long acuminate from an ovate base carinate, margins entire, strongly revolute from base to apex, smooth or scarcely papillose, costa percurrent, 60μ thick at base, cells subuniform, quadrate-rotund, 6μ wide, a few at base near costa short-rectangular, 9μ long. Perichaetial leaves similar to the stem leaves, yet a little more longly acuminate. Seta reddish, about 10 mm. long. Capsule erect, oblong or subcylindric, 2.5-3 mm. long together with the conic-rostrate operculum. Calyptra covering the upper half of the capsule. Annulus distinct, persistant. Peristome purple, 0.6 mm. high, teeth semi-twisted. Spores smooth, pale, 8-9μ in diameter. Plate XVIII.

California: Stanford University, on stones bordering flower beds (C. F. Baker, 1901); foothills near Palo Alto, on ground (C. F. Baker, 1902); Alma, Santa Clara Co., on bowlder (C. F. Baker, 1902): Soldier's Home, Los Angeles Co. (Dr. Hasse, 1902: herb. C. F. Baker).

A species belonging to the perplexing group of *B. fallax* Hedw. differing from this species by its smooth or very slightly papillose leaves with margins more broadly revolute, and by its shorter, slightly twisted peristome. From *B. virescens* Lesq. it is distinguishable by its shorter leaves, broader at base, its almost uniform areolation, and its lower cells small, quadrate, chlorophyllose, with thinner walls.

GRIMMIA COGNATA Card. & Thér.

Apparently dioicous, rather densely caespitose, yellowish green above, fuscous below. Stems usually denuded at base, ascending or arcuate, sparingly branched, 2-5 cm. long. Leaves erect-flexuous when dry, erect-open when moist, 2.5-3.5 mm. by 0.7 mm., lanceolate, gradually and longly acuminate, carinate, extending into an entire hyaline hair point, margins somewhat revolute on one side, areolation opaque, lower cells linear, 30-40μ by 6-8μ with walls strongly sinuate and incrassate. the other cells roundish-quadrate or short rectangular, bistratose in the upper portion. Other characters unknown. Plate XIX.

Colorado: Along the Cogwheel Railway to Pike's Peak, alt. 2100-3000m. (J. M. Holzinger, 1896).

Closely allied to *D. trichophylla* Grev., of which it may be considered as a subspecies; differing from it in its more robust size, recalling that of *G. elatior* Br. Eur., its stems naked below, and the basal areolation with much thicker and more sinuate walls.

To be continued.

IMPORTANT

The second edition of MOSSES WITH A HAND-LENS describes 168 species of Mosses and 51 species of Hepatics, nearly every one being illustrated.

Instead of being a book of 150 pages as at first advertised it will contain 200 pages. There are over 40 full page plates. The additional cost necessitates an increase in price and the price on all orders received after *April 1st* will be $1.75, postpaid.

NOW is the time to subscribe for MOSSES WITH HAND-LENS AND MICROSCOPE. On and after the date of issue of the fifth part the price will become $1.25 per part. So many demands for a complete manual of the Mosses of the Northeastern U. S. that a supplement to MOSSES WITH A HAND-LENS AND MICROSCOPE will be issued immediately upon its completion. The supplement will contain full keys to all species and descriptions and illustrations of all species not included in the main work.

SEND FOR SAMPLE PAGES

A. J. GROUT ❦ **360 Lenox Road** ❦ **Brooklyn, N. Y.**

VOLUME VIII.   NUMBER 3

MAY, 1905

THE BRYOLOGIST

AN ILLUSTRATED BIMONTHLY DEVOTED TO

NORTH AMERICAN MOSSES

HEPATICS AND LICHENS

EDITORS:

ABEL JOEL GROUT and ANNIE MORRILL SMITH

CONTENTS

Entered at the Post Office at Brooklyn, N. Y., April 2, 1900, as second class or mail
matter, under Act of March 3, 1879.

Published by the Editors, 78 Orange St., Brooklyn, N. Y., U. S. A.

THE BRYOLOGIST

BIMONTHLY JOURNAL DEVOTED TO THE STUDY OF NORTH AMERICAN MOSSES
HEPATICS AND LICHENS.

ALSO OFFICIAL ORGAN
OF THE SULLIVANT MOSS CHAPTER OF THE AGASSIZ ASSOCIATION.

Subscription Price, $1.00 a year. 20c. a copy. Four issues 1898, 35c. Four issues 1899, 35c.
Together, eight issues, 50c. Four issues 1900, 50c. Four issues 1901, 50c. Four Vols. $1.50
Six issues 1902, $1.00. Six issues 1903, $1.00. Six issues 1904, $1.00.

*Short articles and notes on mosses solicited from all students of the mosses. Address manu-
script to A. J. Grout, Boys' High School, Brooklyn, N. Y. Address all inquiries and subscriptions to Mrs. Annie Morrill Smith, 78 Orange Street, Brooklyn, N. Y. For advertising space address Mrs. Smith. Check, except N. Y. City, MUST contain 10 cents extra
for Clearing House charges.*

Copyrighted 1905, by Annie Morrill Smith.

THE SULLIVANT MOSS CHAPTER.

President, Mr. E. B. Chamberlain, Washington, D.C. Vice-President, Mrs. C.W. Harris,
125 St. Marks Avenue, Brooklyn, N. Y. Secretary, Miss Mary F. Miller, 1109 M Street,
Washington, D. C. Treasurer, Mrs. Smith, 78 Orange Street, Brooklyn, N. Y.
Dues $1.10 a year, this includes a subscription to THE BRYOLOGIST.
All interested in the study of Mosses, Hepatics, and Lichens by correspondence are
invited to join. Send dues direct to the Treasurer. For further information address the
Secretary.

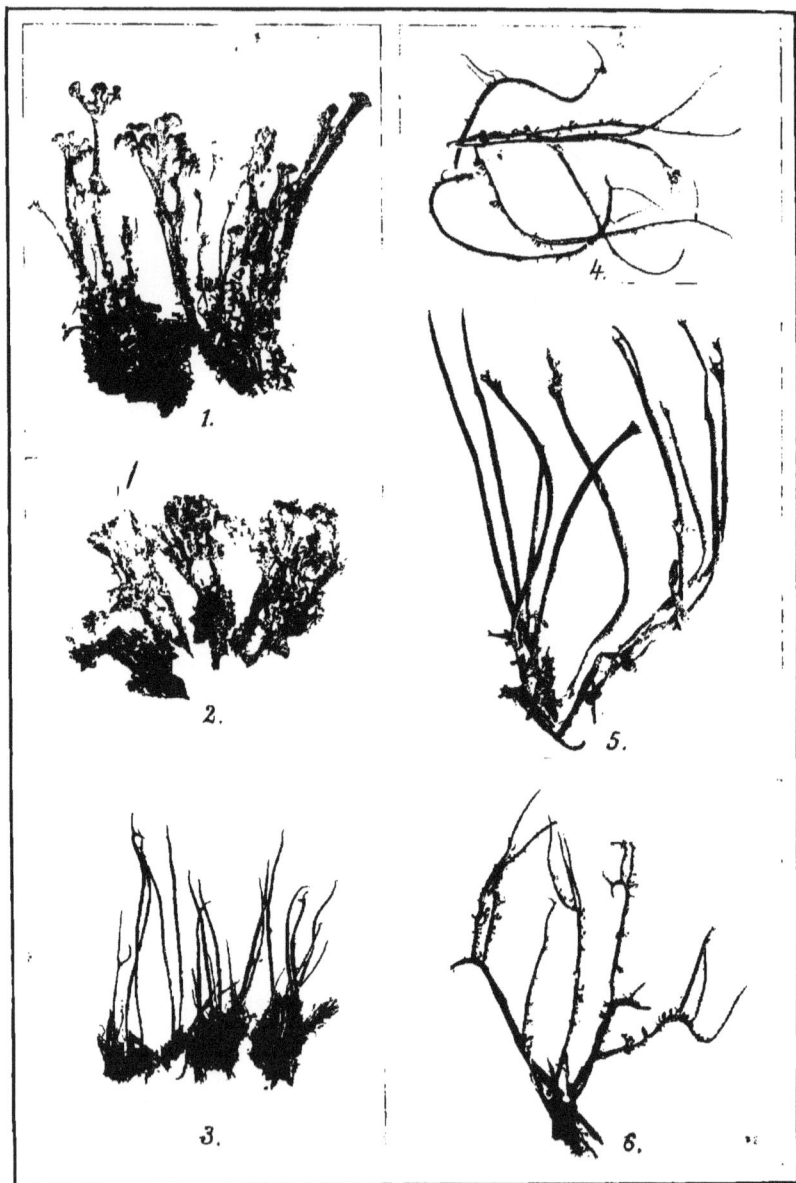

PLATE IV. Fig. 1. *Cladonia gracilis* var. *dilacerata*. Fig. 2. Var. *anthocephala*. Fig. 3. Var. *chordalis*. Fig. 4. Var. *aspera*. Fig. 5. Var. *elongata* Fig. 6. Var. *laontera*. All slightly reduced.

THE BRYOLOGIST.

Vol. VIII.   MAY, 1905.   No. 3.

FURTHER NOTES ON CLADONIAS.   V.
Cladonia gracilis.

BRUCE FINK.

As promised in the last paper of this series we will consider in the present paper *Cladonia gracilis*. The species has been greatly abused in being subjected to "the splitting process" by European workers, but Wainio has succeeded in bringing order out of the chaos of names, and one who has learned to use his Monograph finds little trouble in applying his revision to our American forms of the species. The present writer thought years ago that *Cladonia gracilis* was the most difficult of all our Cladonias, but further acquaintance with *Cladonia fimbriata* gives that species first place as a difficult one. And it now appears plain enough that much of the difficulty with *Cladonia gracilis* was really due to an attempt to follow Tuckerman, who included *Cladonia verticillata*, disposed of in our last paper, with the present species. Then, too, *Cladonia gracilis symphycarpia* Tuck. has been parceled out by Wainio to *Cladonia subcariosa* and *Cladonia cariosa*. Tuckerman gave his variety this description, "cups obsolete from the first, apothecia confluent," and this was wholly inadequate so that no one could conceive what he meant without seeing the specimens. This Wainio has done and has no doubt placed them where they belong. Indeed. it is apparent enough now, after Wainio has done the work, that Tuckerman's brief diagnosis would apply to a form of *Cladonia cariosa* or *Cladonia subcariosa* as well as to one of the species treated in this paper.

Is it any wonder that we could never understand *Cladonia gracilis* while attempting to follow Tuckerman in placing forms of at least four species here ? And, though our forms of *Cladonia cariosa* and closely related species should not be confused with *Cladonia gracilis*, we will consider these plants in the next paper of this series and attempt to remove whatever confusion exists by as good descriptions and figures as we can produce.

In treating the present species, the writer considers himself exceedingly fortunate in being able to see the specimens collected by Mr. G. K. Merrill, on Mount Washington, N. H., during the last summer, and in being able to present figures from photographs taken by Mr. Merrill, from specimens collected in this best known American collecting ground for the species. Indeed, but for the keen-eyed work of this collector, we should have to present figures made partly from European specimens, and we are under great obligations for both the specimens and the photographs.

Before passing to the descriptions of the various forms of the species, it should be stated that the eastern forms are as a rule longer and more slen-

The March BRYOLOGIST was issued March 7th, 1904.

der than the western. This appears in comparing the figure of *Cladonia gracilis dilacerata* presented with this paper and collected by Mr. Merrill in Knox County, Maine, with that of *Cladonia gracilis dilatata* figured in the last paper of this series, and collected on Isle Royale, in Lake Superior, by Mr. Edward L. Harper. And it is again shown in comparing the form *anthocephala*, collected by the writer in Minnesota, with any of the other figures presented in this paper. Likewise, there is recorded in this paper a specimen of *Cladonia gracilis elongata* from Montana, collected by L. H. Pammel, and this also is more inclined to be shorter and runs into ordinary forms of the species. Then to the eastward we get other elongated varieties than *elongata*, as recorded below, and these elongated forms seem to predominate, at least on Mount Washington.

Also, Tuckerman states that the plants are paler in lower altitudes as in lower portions of Maine, Massachusetts and California. Our forms from Minnesota are paler than those collected recently in New England by Merrill, but it appears also that forms of *Cladonia amaurocraea*, quite elongated and with cups, have been frequently placed under *Cladonia gracilis*. Such forms are thus disposed of in the Tuckerman herbarium at Harvard, and a recent specimen collected by W. C. Farlow, on Mount Washington, and named *Cladonia gracilis*, is surely the elongated *Cladonia amaurocraea* and considerably paler than forms of the present species. Also *Cladonia gracilis chordalis*, "Lichenes Boreali-Americani," no. 272, seems to be this same *Cladonia amaurocraea*.

CLADONIA GRACILIS (L.) Willd. Fl. Berol. 363. 1787.

Primary thallus usually persistent, composed of irregularly laciniate or crenate. somewhat flat, involute or convolute, ascending. clustered or scattered squamules, which are somewhat incrassate and middling sized, 2-5 mm. long and nearly as wide, sea-green varying to olivaceous above, and white below or brownish toward the base. Podetia arising from the surface of the squamules, 10-75 mm. long and .3-5.5 mm. in diameter, cylindrical and cupless, or more or less narrowly trumpet-shaped and scyphiform, commonly in larger or smaller clusters, erect or ascending, the cortex subcontinuous or composed of contiguous or scattered areoles. rarely squamulose, the decorticate portions between the areoles sometimes granulate sorediate, variously sea-green, olivaceous, or even reddish-brown, the decorticate portions white, sometimes dying below and the dead portions becoming dark colored, simple or more or less branched, the sides sometimes more or less rimose or perforate. Cups .75-6 mm. in diameter, abruptly or gradually dilated, regular or irregular, shallow or deep, the margins dentate or proliferate (rarely proliferate from the centre ?), the ranks from one to five, the lowest rank from 10-70 mm. long, and when four or five ranks the whole podetium longer than stated above. Apothecia medium sized, 1-4.5 mm. in diameter, usually lobate-conglomerate and sometimes perforate, commonly borne on short pedicels. which frequently arise singly or in clusters from the margins of the cups, thinly margined or more commonly convex and immarginate, pale or darker brown. Hypothecium pale. Hymenium pale below and brownish above. Paraphyses rarely branched, thickened and brownish toward the apex Asci cylindrico-clavate

Widely distributed over North America in one form or another, and specimens not yet assigned to any of the varieties have been seen by me from Massachusetts, Mount Washington, Ontario, Iowa, Minnesota, Illinois, and Alaska. This indicated a wide distribution throughout northern North America, corresponding thus with the views of Tuckerman and Wainio No specimens of the species are known to the writer from the southern half of the United States. The plant is known in all of the grand divisions.

CLADONIA GRACILIS DILATATA (Hoffm.) Wainio Mon. Clad. Univ. 2:87. 1894.

Podetia frequently stouter, 1-5 mm. in diameter, scarcely exceeding 50 mm. in length and usually much shorter, destitute of squamules or rarely squamulose toward the base, neither granulose-sorediate nor decorticate, cortex continuous or sometimes areolate, cups quite dilated and subregular, the ranks rather short

Tuckerman's var. *hybrida* Schaer. is this and the next as he says, "podetia *often* beset with squamules." He states that his variety is found, wherever var. *elongata* occurs, which is true and more. For the present variety and the next occurs at lower altitudes than *elongata* to the south, and also farther south. I have examined material of my own collecting of the present variety from Iowa, Minnesota and Massachusetts, and by others as follows: New York (E. A. Burt), White Mountains (H. Willey), Montana (R. S Williams), Columbia River and Nova Scotia (J. Macoun), Newfoundland (A. C. Waghorne), and Alaska (Kincaid?). Macoun reports var. *hybrida* as very common throughout British America, and his specimens are no doubt nearly all this common variety, rather than the next much rarer form. Miss Clara E. Cummings also reports the present variety from Alaska. Reported from all the grand divisions. (See last paper of this series for figure of the variety, with squamules at base and near the next below).

CLADONIA GRACILLIS DILACERATA Flk. Clad. Comm. 37. 1828.

Differing from the last in that the podetia are always squamulose elsewhere as well as at the base, and often conspicuously so at the top of the podetia among the apothecia. The cups also are rather more inclined to irregular forms than in the last. *Cladonia gracilis anthocephala* Flk. Clad. Comm. 37. 1828, Wainio considers a form of the present with numerous squamules at the top of the podetia. Plate IV. Figs. 1 and 2.

Wainio credits this variety from Greenland and states that Tuck. Lich. Amer. Exsic. no. 27 is this in part. He has determined the ordinary form, and the one with squamules numerous upward, from northern Minnesota. The variety is evidently rare in the United States, and the writer has seen it only from Minnesota, of his own collecting, and from Maine (G. K. Merrill) and Yellowstone Park (Aven Nelson). Much rarer and only known, outside North America, in Europe.

CLADONIA GRACILIS CHORDALIS (Flk.) Schaer. Lich. Helv. Spic. 32. 1823.

Podetia quite slender, subulate, or tubaeform and scyphiform, destitute of squamules or rarely and very sparsely squamulose, neither granulose nor decorticate, cortex continuous or areolate. elongated. 30-140 mm. long and only about .1-2 mm. in diameter, sparingly branched, and the branches at least partly subulate, the lowest rank long. Cups narrow, only about 2-5 mm. in diameter, regular, or sometimes oblique or irregular, and often radiate or proliferate. Plate IV. Fig. 3.

Wainio credits specimens from Great Bear Lake and Greenland, and Tuckerman barely mentions the variety under *Cladonia gracilis elongata* as a form of that variety. Miss Cummings found the variety to be common in the material examined from Alaska. I have determined it from New-foundland (A. C. Waghorne) and from Mount Washington (G. K. Merrill). Also from Mt. Ranier region, Washington, 1,500 ft. (T. C. Frye). This variety is known in all of the grand divisions.

CLADONIA GRACILIS ASPERA Flk. Clad. Comm. 40. 1828.

Podetia and cups much as in the last, but the podetia usually shorter, never exceeding 100 mm. and always squamulose. also rather more slender as a whole, and rarely branching more freely. Plate IV. Fig. 4.

Wainio reports this variety from Miquelon Island, and I have determined it from Mount Washington for G. K. Merrill. Nothing more is known of North American distribution, but the plant is recognized in Europe, Asia and Africa.

The worth of the variety may be questioned as it seems to stand in nearly the same relation to var. *chordalis* as the form *laontera* noted below does to *Cladonia gracilis elongata*. However, there are other slight differ-ences than the presence or absence of squamules, at least in some instances. and we let the variety stand rather than depart from Wainio's view with-out sufficient study. Varieties *chordalis* and *aspera* stand in somewhat the same relationship to each other as varieties *dilatata* and *dilacerata*, but better acquaintance has emphasized the difference between the last two.

CLADONIA GRACILIS ELONGATA (Jacq.) Flk. Clad. Comm. 38. 1828.

Podetia not surpassing the extreme length given for var. *chordalis*. but on the whole longer, averaging about 75 mm. long and *chordalis* only 50 mm., likewise plainly stouter, 1-5 mm. in diameter, branching rather less frequent than in the last two varieties, but sometimes quite freely branched. destitute of squamules (except in the little-known form *laontera*), cortex continuous or areolate, subulate or scyphiform. Cups also as in last two varieties, except that they are larger, 2-6 mm. in diameter, or rarely even 12 mm. Plate IV. Fig. 5.

Wainio credits the plant from Greenland, Labrador, Kotzebue's Sound, Port Clarence, Vancouver Island, British America, White Mountains and western United States. John Macoun lists it from all parts of British America, and Miss Cummings reports it as common in Alaska. I have examined it from Mount Washington (G. K. Merrill, 4,000-6,000 ft.), Knox County, Maine (also by Merrill), the Adirondack Mountains (Carolyn W. Har-ris) and from Warm Springs in Montana at 5,000 ft. alt. (L. H. Pammel). The squamulose form, *Cladonia gracilis laontera* (Del.) Arn. Rehm. Clad

Exsic. no. 261, Plate IV. Fig. 6. was also sent me from Mount Washington by G. K. Merrill. This form is little known even in Europe, and is not regarded worthy of varietal rank, Tuckerman doubtless knew of it, for he speaks of the squamules sometimes occurring in the present variety. We figure the form *laontera* as well as the usual expression of the variety. The variety is known in all the grand divisions except Australia, commonly in mountains.

The various forms of *Cladonia gracilis* commonly occur on earth, on horizontal rocks covered with a small amount of humus, or on rotting wood. The plants are usually found in forests, preferring shade, and are often among mosses. The last variety descends to the tundras in northern regions and there often occurs in sunny and windy places.　　　　　Grinnell, Iowa.

- - - -

PHAENOLOGICAL OBSERVATIONS ON MOSSES.

H. WILHELM ARNELL.

In the BRYOLOGIST there appeared, in the year 1904, p. 35-36, a note on "The Fruiting Season of the Hair-cap Moss," by Phebe M. Towle and Anna E. Gilbert, that interested me much as it touched upon a subject which I myself have studied. A rather long time ago I studied the seasons of blooming and fruiting of the Scandinavian mosses (musci veri). The results of my researches were summed up in a paper, " De Scandinaviska Löfmossornas Kalendarium (Upsala Universitets Arsbkrift, 1875) in which the seasons of blooming and fruiting of all Scandinavian mosses known at that time were indicated as accurately as the material that was accessible permitted me to solve these questions. Since that time there has, as far as I know, nothing been published on the phaenological relations of the mosses till the paper of A. Grimme, " Ueber die Blüthezeit Deutscher Laubmoose und die Entwickelung ihrer Sporogone " appeared in Hedwigia, 1903.

I will now tell the results to which I have come with regard to the Hair-cap mosses. The species of Polytrichum, I found phaenologically to constitute two different groups. The species belonging to Ymnitrichum (Neck) Lindb., namely *P. urnigerum*, *P. nanum* and *P. subrotundum*, bloom at an early season at Upsala, in May or the first week of June, and the lids of the sporogones are therefore cast in the winter or early spring, at Upsala in December to April. These mosses thus require less than a year or seven to ten months for the development of their sporophytes. The majority of the Scandinavian Polytricha, the sections Pterygodon Lindb. and Euleiodon Lindb., bloom early in the summer, at Upsala in June and July, while the sporogones become ripe in July or the first days of August. These species require more than a year or about thirteen months for the development of their fruits. The two species of Polytrichum, *P. juniperinum* and *P. commune*, observed by the Misses Towle and Gilbert, both belong to the latter group of Hair-cap mosses. A comparison of their dates with mine shows a rather great difference, as according to the observations in Vermont, these mosses bloom in April, and the maturing of the spores takes place in August of the following year, which difference probably depends on the different latitudes in which the observations are made, If the Vermont dates are correct, which fact I have no reason to doubt, *P. commune* and *P. juniperinum* in this region require fifteen to sixteen months for the development of their sporophytes.

The interest in the phaenological relations of mosses, of which the note in THE BRYOLOGIST bears evidence, encourages me to renew in a modified form a proposal of phaenological observations on mosses that I made in 1878, in the "Revue Bryologique," which proposal, however, has hitherto led to no results. In phanerogams the phaenological observations are, as is well known, always continued through as long a period of years as possible in order to get fully comparable medium dates in which the variations of single years are eliminated. Desirable as such medium dates may be in mosses, I do not, however, dare to insist upon the carrying on of observations through a period of years in each place, because most bryologists will shrink from undertaking so toilsome a task. Dates from even a single year will be of great value and will open the possibility of interesting comparisons of the phaenological differences of mosses in different regions.

As it is *the beginning* of the blooming and the fructification that have been made the object of phaenological observations in phanerogams, I propose that the same phenomena be observed also in mosses. The beginning of the blooming in mosses is to be considered to occur when one or two archegonia only, in at least two flowers of a species, are opened, while the other archegonia are still closed and uncolored, or, if archegonia are not accessible, when only one or two antheridia in at least two flowers are opened and uncolored or brown, while all the other antheridia are still closed. In determining the blooming time however, if archegonia are at hand, these ought in the first place to be consulted, because we are less subjected to errors when determining the blooming time of a moss by the archegonia than when judging by the antheridia. Especially I warn observers against believing the blooming to have begun in a moss merely because its antheridia appear fully developed, if at the same time no antheridium is opened, as antheridia may often appear fully developed many months before they open and their antherozoides begin to swarm. In getting the date of the first blooming of a moss I have found it the least labor to collect specimens of the species that I desire to examine at different times of the year, for instance every fortnight, and then afterwards to examine all the collected specimens at once.

The fructification is according to my proposal considered to have begun when in a species about five to ten lids are naturally detached from their fruits; in dried specimens the lids are, as every bryologists knows, detached earlier than in nature, on which account dried specimens are less adapted for these researches.

My proposal was limited to only a small number of species. In making the choice of species I tried to get mosses that have a wide distribution, are easily recognized, have a limited season of blooming and fruiting and, as far as possible, represent every different part of the year, etc.   I will now enumerate the mosses chosen by me for observation on account of one or another of the reasons mentioned.   To this enumeration I have added the results to which I myself have come in the midst of Sweden, at Hornoesand, 62° 30′

	HERNOESAND, SWEDEN			GERMANY		
	Bloom	Fruits	Devel.	Bloom	Fruits	Devel.
*1. Acrocladium cuspidatum (L) Lindb.	15. VIII.	7. VII.	11	VIII.	VI.	10
2. Amblystegium cordifolium(Hedw.) DeN.	15. VIII.	7. VII.	11	VII. VIII.	VI. VII.	10-12
3. Anisothecium rubrum (Huds.) Lindb.	20. VI.	10. V.	11			
4. Astrophyllum punctatum (L.) Lindb.	20. VI.	1. VI.	11	IV. V.	IV. VI.	11-14
5. Astrophyllum silvaticum (Lindb.) Lindb.	1. VII.	20. VI.	11	III. VI.	IV. V.	10-14
Barbula unguiculata(Huds.) Hedw.	1. VII.	1. V.	10	VII. VIII.	III. IV.	7- 9
Bartramia pomiformis (L.) Hedw.	15. VII.	1. VII.	11½	V.	V.	12
Catharinea undulata (L.) W. & M.	1. VII.	1. V.	10	V.	II. III.	9-10
Ceratodon purpureus (L.) Brid.	7. VII.	1. VII.	12	VII. VIII.	V. VI.	9-11
Dicranella cerviculata (Hedw.) Schimp.	1. IX.	1. V.	20	IX.	IX.	12
Dicranum fuscescens Turn.	25. VII.	15. X.	15	V.	VIII.	15
" undulatum Ehrh.	25. VII.	15. X.	15	VI. VII.	X.	15-16
Funaria hygrometrica (L.) Sibth.	1. IX.	1. VIII.	11	X.	VII. X.	9-12
Georgia pellucida (L.) Rab.	10. VI.	15. VII.	13	V.	V. VI.	12-13
Grimmia apocarpa (L.) Hedw.	7. VII.	20. V.	10½	V.	III. IV.	10-11
6. Hedwigia albicans (Web.) Lindb.	7. VII.	10. V.	10	VI. VII.	VI.	11-12
7. Hylocomium parietinum(L.) Lindb.	7. VII.	10. V.	10	V.	II. III.	9-10
" proliferum(L.) Lindb.	15. VII.	10 VI.	11	V. VI.	IV. V.	10-12
" triquetrum (L.)B.& S.	7. VII.	10. V.	10	V.	III. IV.	10-11
8. Hypnum plumosum Huds.	5. VIII.	10. V.	9			
" strigosum Hoffm.	25. V.	20. V.	12	VII.	III.	9
Philonotis fontana (L) Brid.	1. VIII.	15. VII.	11½	V. VI.	V.	11-12
Plagiothecium denticulatum (L.) B. & S.	1. IX.	1. IX.	12	VII. VIII.	VII.	11-12
9. Pohlia cruda (L.) Lindb.	1. VII.	15. VII.	12½	IV.	V.	13
Polytrichum commune (L.)	1. VII.	1. VIII.	13	V. VI.	VII. VIII.	13-15
" pilosum Neck.	1. VII.	15 VII.	12½	IV. V.	VI. VII.	13-15
Ptilium crista-castrensis (L.) DeN.	15. VIII.	1. V.	20			
10. Sphaerocaphalus palustris (L.) Lindb.	20. VI.	25. VII.	13	V.	VI.	13
11. Sterodon cupressiforme (L.) Brid.	10. VI.	10. VI.	11	IV. V.	III.	8-11
12. " incurvatus (Schrad) Mitt.	5. VIII.	1. VIII.	12	VII. VIII.	VII.	11-12
13. " polyanthos (Schreb) Mitt.	10. VI.	10. V.	11	VIII. XI.	II. III.	15-19
Tortula ruralis (L.) Ehrh.	25. VII.	7. VII.	11½	IV. V.	V. VI.	12-14
14. " truncatula (L.) Lindb.	25. VII.	1. V.	9	VI. VII.	I. II.	6- 8

* Note. Current synonyms of the above mentioned mosses are : 1. *Hypnum cuspidatum*, 2. *Hypnum cordifolium*, 3. *Dicranella varia*, 4. *Mnium punctatum*, 5. *Mnium cuspidatum*, 6. *Hedwigia ciliata*, 7. *Hypnum Schreberi*, 8. *Brachythecium salebrosum*, 9. *Webera cruda*, 10 *Aulacomnium palustre*, 11 *Hypnum cupressiforme*, 12. *Hypnum incurvatum*, 13. *Pylaisia polyantha*, 14. *Pottia truncatula*.

N. Lat., and A. Grimme, in Germany, with regard to the seasons of the blooming (Bloom.) and the maturation of the fruits (Fruits.) of mosses in question, as also with regard to the time (expressed by the number of months) which these mosses require for the development of their fruits (Devel). The days are expressed by the usual figures, the months by Roman numerals, as I (January), II (February), etc.

From the above table it will be seen that the results to which Grimme and I have come are similar, if the differences due to latitude are considered. With regard to two mosses, *Dicranella cerviculata* and *Stereodon polyanthos*, our results show greater difference; acording to my experience in Sweden, *Dicranella cerviculata* requires about twenty months and *Stereodon polyanthos* eleven months for the development of the sporophytes, while these mosses require in Germany, according to Grimme, the former species only twelve months and the latter species fifteen to nineteen months for the same purpose. The dates contained in the table afford material for many interesting comparisons, and I hope they will induce some readers of THE BRYOLOGIST to examine the phaenological relations of the same mosses in their locality and by publishing the result procure more material for comparison.   Upsala, Sweden.

NOTES ON THE FRUITING SEASON OF CATHARINEA.

PHEBE M. TOWLE.

During the season of 1903 some attention was given to the fruiting season of the Hair-cap moss, and the report was published in THE BRYOLOGIST of March, 1904.

Observations for the sake of verification were repeated on the Hair-caps in 1904, with the same conclusions as were noted for 1903.

This same line of observations was followed during the season of 1904 with reference to *Catharinea undulata*.

This Catharinea is abundant about Burlington. The stations visited were in the ravine north of Colchester avenue and along the brook which runs through the woods of Fair Holt. Another station was along the bank of the Missisquoi River.

On the first trip afield, the last of March, Catharinea was found with the capsules shedding their spores freely when disturbed. This continued through April, the spores coming out in clouds.

On April 29th Catharinea rosettes were conspicuous. The antheridia were green in color, but had a narrow, clear space at the top. None of them were ready to shed their contents.

May 3d, Catharinea tufts contained archegonia of different heights but none were open.

May 7th, the antheridia were discharging sperm mother-cells, and three of the five archegonia of a female plant were open. There were five archegonia in each of the Catharinea tufts examined. When the archegonia were open they presented a well spread border around the top of the tube.

The tube and the egg cell cavity were clearly seen. They were light and bright, doubtless due to the refractive quality of the cytoplasm.

May 13th, an archegonium was observed in which a drop of dark fluid was passing slowly down the tube. It reached the egg cell. The egg cell changed in appearance, dividing, then uniting and finishing for the time of observation in a bright cell with a small dark spot on the upper side. This may have been the downward passage of an antherozoid. It was not clear enough for a positive statement, but it suggests something to be looked for about the middle of May.

June 14th, in a tuft examined one archegonium had a sheath around its base. The cover glass was pressed and the embryo sporophyte came out at the base. The other four were nearly as tall but had no sheaths.

July 14th, the Catharinea sporophytes were about one-half inch above the leaves. They looked like slender green stems. One week later the capsules began to show, and by August 1st the calyptras were seen.

October 1st, the sporophytes were of full height but the capsules were not completely filled out. December 1st, these seemed full grown but there was no suggestion of their capsules opening.

Bringing forward the first observation of the year it may be safe to conclude that these capsules will discharge their spores next March and April.

This Catharinea therefore developes more rapidly than the Hair-cap. For Catharinea the time from the maturity of the antheridia and archegonia and their contents, to the maturity of the sporophytes is about eleven months, while for the Hair-cap it is seventeen months.

Botanical Laboratory, University of Vermont, Jan. 18, 1905.

- - - - - - -

LICHENOLOGY FOR BEGINNERS.

FREDERICK LEROY SARGENT.

In popular language we hear lichens usually spoken of as "mosses," a name which, unfortunately, belongs by right to little leafy plants very different from lichens, and has, moreover, been applied to so many other kinds as to have become sadly indefinite. On the other hand, the name *lichen*—which was given to this group about two hundred years ago by the great botanist Tournefort—has the advantage of being used for these plants alone, and is therefore much to be preferred.

It is usually an easy matter to distinguish a lichen at sight. In the first place, the absence of anything like leaves regularly arranged upon a stem, at once separates them from the true mosses and higher plants. Then there is the peculiarity that, although a great variety of colors are exhibited, there is one—the bright green of foliage or grass—which, under ordinary circumstances, they never assume; and in this respect they differ from certain liverworts which in form they sometimes resemble. Although they often grow where there is moisture, lichens are not aquatic, and thus they differ from *Algae*, which, with few exceptions, grow only in the water. Finally, the fact that lichens are air-plants—*i. e.*, derive all their nourishment from the air—serves to separate them from fungi, which require for food either living animals or plants, or else their decomposing remains. In short, given a plant which is leafless, of some other color than grass-green, aerial, and dependent merely for mechanical support on the object to which it is attached, and we may be reasonably sure that our plant is a lichen. For final decision, in certain cases, a microscopical examination of the internal structure is necessary; but to this matter we shall return.

The extreme simplicity of their needs, and their indifference, for the most part, to cold, heat, drought, and other rigors of climate, make it possible for lichens to grow in a great variety of situations, which, but for them, would be scarcely occupied by vegetation at all. Wherever a tolerably permanent foothold is offered, with a chance of occasional moistening by dew or rain and an exposure to pure air and daylight, there one may expect to find lichens. They grow on rocks or stones, morter, brick, the bark of trees, perennial or evergreen leaves, old wood-work, stumps, and barren earth, and occasionally on bones, leather, or bits of crockery exposed to the weather. Their distribution over the face of the earth extends from sea-shore to the highest mountain-tops, and from the equator to as near the poles as man has been able to penetrate.

Although as a group they are distributed so much more widely than most other plants, and appear in general remarkably indifferent to their surroundings, nevertheless there are not a few species of lichens which show strong preferences for certain situations, and will thrive only under certain conditions. Thus, there are forms which occur only in northern regions or at high elevations, or which produce fruit only in such localities. Similarly, temperate and tropical regions have their peculiar forms, and likewise sub-Alpine and lower levels. Certain forms of lichens are found only on bark, certain others on rocks alone, and others, again, only on the earth. There is even with some species a preference for a particular sort of bark, or rock or soil. Thus, peaty, clayey, and sandy soils have different species growing on them; and calcareous, arenaceous, and granitoid rocks each have their own forms, which are rarely found elsewhere. Smooth-barked trees harbor lichens different from those on rough-barked trees. Even particular species of trees have their own peculiar forms of lichens, which seldom or never grow elsewhere. This is true of the cinchonas and certain other trees from which medicinal barks are obtained, and pharmacognocists have taken advantage of this fact to aid them in distinguishing the different varieties of bark imported.

The amount of moisture to be obtained and the degree of exposure to daylight are also important factors in deciding the dwelling place of many lichens. It is a fact well known to woodsmen that lichens in this part of the world are apt to grow most luxuriantly on the northern or northwestern side of tree trunks. This circumstance is often helpful in finding one's way through a forest. The reason why lichens choose one side by preference appears to be that, being less exposed to the direct rays of the sun, it remains moist somewhat longer after a rain; but exceptions abound.

Unlike most other plants, lichens are found in as perfect condition one season as another. Soon after a rain is a good time to gather them, as they are then more easily handled than when dry and brittle.

No very elaborate collecting outfit is necessary. A good sharp jack knife and something to contain the specimens are the most important requisites. For holding the different kinds as collected, the writer finds paper bags such as grocers use, particularly convenient. They take up so little room that a supply may be carried in the pocket and so be always handy for the collection of any good specimen that may turn up unexpectedly in one's walks. When one expects to gather more specimens than might be easily carried in one's pocket. a light basket is to be recommended. In order to obtain specimens which adhere firmly to rocks, bark and the like, it is necessary to detach portions of the substance upon which they grow. For this purpose it is well to have a cold chisel and a small sharp hatchet which may be used either as a hammer in connection with the chisel or to chop off slices of hard bark or wood. Only experience can teach the collector how to obtain good rock specimens, but a little practice will show how blows may be most effectively directed. After a fragment has been detached it is often possible to trim off superfluous material, and for obvious reasons it is best when possible to do this at the time. The specimen should always include enough of the margin of the lichen to show its character. To prevent the heavier specimens from injuring the more delicate ones, it is well to put the latter in a box by themselves. If one is collecting in a dry time a small sponge and some water is convenient to have for moistening such specimens as would be otherwise too brittle to be easily detached without injury.

While collecting, the conditions under which a given sort are growing, should be carefully noticed, and a record of this, together with the locality, should be made on the bag or wrapper which is to contain the specimen. For example such a note might read: "On trunk or branches of apple trees. Cambridge, Mass., March, '91." The date, however, is not so necessary in the case of lichens as with most other plants.

Whenever possible it is important to obtain specimens in fruit (the form of which will be described later). Certain species may never produce fruit in the student's locality. but in such cases their study is best pursued in the light of knowledge previously gained from perfect specimens of allied forms.

Some species have the fruit so small and inconspicuous as scarcely to be seen with the naked eye. It is well, therefore, to have at hand a pocket lens magnifying about ten to twenty diameters, to scrutinize such minute structures.

Soon after the specimens have been brought home is the best time to prepare for the herbarium such as need preparation. The slices of bark will curl unless dried under pressure. If they are already curled and dry, soaking in cold water will render them flexible, after which they may be put in press. Warm water should never be used to moisten lichens as it discolors them. Specimens which are shrubby or of considerable thickness may be flattened under light pressure between folds of a newspaper. With such species as grow in close contact with the earth, it is often desirable to preserve intact a thin layer of the soil, so as to hold the plant together. This is accomplished by a coating of mucilage applied to the underside while the earth is moist.

There are several methods in use among students of lichens, for preserving and arranging their specimens. Different circumstances and requirements make different ways preferable. A favorite way to keep the specimens is in pockets, which are made as follows: Fold a piece of paper (7 by 9 inches is a convenient size) so the under part shall project about three-quarters of an inch beyond the upper; fold the projecting part over the other to make the top of the pocket; turn the pocket over and fold back the right and left edges for about three-quarters of an inch, and it is finished. Such pockets, if of uniform size, may be arranged in a box or tray like the cards of a card catalogue. When thus used, writing the name of the species at the top of the pocket, facilitates reference. This method of keeping specimens affords a way of making a good working collection with small outlay of time and labor. If it is desired to incorporate the specimens in a general herbarium, the pockets may be attached to the regular species-sheets by means of small ribbon pins, or with a spot of glue at the back. The chief objection to using pockets is that the specimens are not displayed. To accomplish this, specimens are glued onto pieces of rather stiff paper of convenient size (room being allowed for labeling), and then these pieces of paper are pinned to the herbarium sheets. In thus mounting lichens which have a distinct under side it is important that the character of this as well as the upper side be exhibited. Keeping specimens in scrap-books is hardly to be recommended, unless for a very small collection, since this method does not facilitate a systematic arrangement and interpolation of additions.

Rock specimens which are too bulky to be incorporated with the others may be kept like minerals in trays or in a cabinet. An approved way for large collections is to have pasteboard boxes made of such a size as to go with the rest of the collection in the herbarium case.

All specimens should be labeled with the botanical name of the species, the locality, habitat, and date of collection (if desired) together with the name of the collector (if other than the owner).

Having collected as many as possible of the common species of the neighborhood and prepared them conveniently for examination, the student is in position to begin an examination of their form, structure and classification.

A good deal may be learned about them without other aid than the simple microscope already recommended; but for thorough study of lichens, a good compound microscope, magnifying three to four hundred diameters, is essential. We shall consider, however, at this time chiefly such features as involve in their study comparatively simple microscopical manipulations.

Cambridge, Mass.

(To be continued.)

NOTES ON NOMENCLATURE. V.

ELIZABETH G. BRITTON.

Fissidens decipiens has been giving a great deal of trouble, and masquerading under various names indicating that it is doubtful and deceitful, as well as possessed of other well-marked characters. Its synonymy seems to be as follows:

Fissidens dubius Beauv. Prod. 57. 1805?
Skitophyllum marginatum La Pyl. Desv. Journ. Bot. 4:163. 1814.
Skitophyllum adiantoides marginatus La Pyl. 2. c. 4:164. 1814.
F. adiantoides marginatus Brid. Bryol. Univ. 2:704. 1827.
F. cristatus Wils. Hook. Journ, Bot. 9:294. 1857?
F. decipiens De Not. Cronaca briol. ital. 2:98. 1866.
F. rupestris Wils. Musci Brit. no. 311, Jäg. Enum. Fiss. 25. 1869.
F. Floridanus L. & J. Proc. Am. Acad. 14:137. 1879. Man. 83. 1884.

Of the names given above, the first, second and third were founded on the same specimens collected in America by Beauvois, but without record of the type locality. The specimens were also sterile, and La Pylaie states that they do not appear to differ from *Fissidens adiantoides*, except in their shorter stems, and the remarkably hyaline border of their leaves, which are otherwise the same, having the serrate apex. He referred them to this species as a variety, but cites the name given by Beauvois and his description. When Dr. Barnes was studying the genus he made an effort to see the types, but failed to find them: the cotypes may exist at Paris in the Herbarium of La Pylaie.

Fissidens cristatus was described from specimens collected in the Khasia Mts. of India at an altitude of 5,000 ft.,and the specimens have strongly recurved leaves, They appear to be somewhat different, and may prove to be a good species. They are larger and coarser than *F. decipiens* and the leaves have larger cells.

Fissidens decipiens was described from Italian specimens, of which there are authentic duplicates from the type locality in the Jaeger herbarium. In studying some specimens from Florida it struck me that the descriptions of *F. decipiens* and *F. Floridanus* did not differ materially, so I wrote to Kew asking for comparisons. Both Mr. C. H. Wright and Mr. E. S. Salmon agree that they are the same species. *Fissidens Floridanus* has not been known in America for twenty years, and there are no specimens preserved in the James Herbarium, but authentic specimens were sent to Schimper and are now at Kew. This disposes of another superfluous name, and renders it more than probable that the oldest name belonging to this species is *Fissidens dubius*, Beauv.   New York Botanical Garden.

NOTES ON SOME NORTH AMERICAN MOSSES. II.

JULES CARDOT.

GRIMMIA LAMELLOSA C. Müll, in Bot, Zeit,, 318, 1854.
Limpricht cites erroneously this name as a synonym for *G. alpestris*

Schleich. From the examination of an original specimen of *G. lamellosa* collected by J. Lange at the Lake Espingo, Pryenees, it appears clearly that this moss is the same that Limpricht thirty-five years later described as *G. subsulcata* sp. nova (Laubmoose, I. 757). It has been collected in Montana by Holzinger, and in Idaho by Leiberg and Sandberg. (See Bot. Gaz. XXX, 18, and BRYLOLGIST, V, 14).

PAPILLARIA PENDULA R. & C.

A synonym of this species is *Neckera capilliramea* C. Müll. from Java. I have also specimens from China, Japan and Formosa; the Japanese specimens (Faurie, ser. 2, Nos. 1147, 1182, 1613) have been related to *Papillaria Wallichii* (DeC.) R. & C., a somewhat doubtful species from Nepaul, Java and Sumatra, I have not been able to see the type of *Neckera Wallichii* DeC. (*Hypnum* Brid., *Meteorium* Mitt.) from Nepaul, but the *P. Wallichii* from the Malayan Archipelago, although very near to *P. pendula*, differs from it by its stronger habit, its patent subcompressed branches, its broader leaves and its narrower and more papillose areolation.

The synonomy of *Papillaria pendula* is therefore as follows:

Meteorium ? pendulum Sulliv. Musci and Hep. of the U. S. 681 (81) 1856.

Neckera capilliramea C. Müll., in Bot. Zeit. 237, 1859.

Meteorium pendulum Sulliv. Icon. Musc. 117, t. 73, 1864. L. & J. Man. 286, 1884.

Papillaria capilliramea Jaeg. Ad. 11, 168, 1879.

Papillaria pendula R. & C., in Rev. Bryol. 11, 1893 Musci Amer. Sept. 45, 1893.

Floribundaria capilliramea Fleisch. Musci Arch, Ind. No. 234, 1902.

ANOMODON TOCCOÆ Sulliv. & Lesq.

To my mind this moss is generically distinct from *Anomodon* by its different areolation, formed of well distinct cells nearly uniform, quite smooth, with firm thick and pellucid walls, and also by its stronger nerve which is very flexuous above. The section name *Herpetineuron* C. Müll. is well appropriated, and must be used as generic name, and *A. Toccoæ* must be called *Herpetineuron Toccoæ* Card. Another species of the same genus is *H. Wichuræ* (Brother.) Card., from Japan. As to the synonymy and distribution of *H. Toccoæ*, see Mr. Salmon's paper in Jour. of Bot. XXXIV, 273.

THUIDIUM GLAUCINUM (Mitt.) Borsch. & Lac. Bryol. Jav. 11, 117, t. CCXXII, 1865.

Leskea glaucina Mitt. Musci Ind Orient. 133, 1859.

This is a most interesting addition to the American bryology. Lately in revising the numerous North American specimens of the Tamariscinum group in my herbarium I found some specimens collected in Louisiana by the late Rev. Mr. Langlois, which proved to be different from all the European and North American species, in the stem leaves being ovate-lanceolate, broadly, shortly and obtusely acuminate, and the broader, more concave, obtuse branch leaves. The same characters also distinguish this

plant from the Central American *T. Antillarum* Besch. and *T. miradori-
cum* Jaeg., but I found that it agrees with *T. glaucinum* (Mitt.) Borch. &
Lac., a species widely distributed throughout the Malayan Archipelago,
Formosa, Japan, Ceylon, Assam and Himalaya.

Rev. Mr. Langlois' labels indicate the following localities and dates for
this species: No. 180, Bois de Lafayette, Sept. 25, 1885; No. 264, Forêt de
Lafayette, May 6, 1887; No. 436, Pinieres de Covington, July 18, 1885; No.
860, Abita, St. Tammany Co., Nov. 27. 1891. No. 436 bears some perichae-
tia and young pedicels, and No. 264 numerous male flowers. The form from
Louisiana rather differs from the Asiatic plant by the stronger nerve of the
stem lèaves and by the less denticulate branch leaves, covered on the back
with lower papillæ. It may constitute a var. *ludovicianum* Card.

Charleville, France, Jan. 15, 1905.

ADDITIONS TO THE LICHEN FLORA OF LONG ISLAND.

G. C. Wood.

The following species of Lichens which have been found and determined
by Mrs. Carolyn W. Harris and myself have not hitherto been included in
Jeliffe's or any other check list of Long Island Lichens:

*Cetraria lacunosa, Ach., on trees. Cold Spring Harbor.
*Cetraria aurescens, Tuckerm , on trees. Cold Spring Harbor.
*Evernia furfuracea, (L.) Mann. Cold Spring Harbor.
*Parmelia saxitalis, var. sulcata, Nyl., on trees. Cold Spring Harbor.
*Parmelia perforata (Jaeg.) Ach., on trees. Cold Spring Harbor.
*Physcia obscura (Ehrh.) Nyl. Rocks. Cold Spring Harbor.
Usnea barbata var. florida (Fr.) on trees. Cold Spring Harbor.
Pertusaria pustulata (Ach.) Nyl., on trees. Jamaica.
Pertusaria glomerata (Ach.) Schaer., on trees. Jamaica.
C. caespiticia (Pers.) Fl. Ground. Jamaica.
C. fimbriata (L.) Fr., on earth. Jamaica.
C. decortica. Floerk., on earth. Jamaica.
C. macilenta (Ehrh.) Hoffm., Flushing, Jamaica. Common.
C. cariosa (Ach.) Spreng., on earth. Flushing. Common.
C. fimbriata var. tubaeformis Fr , on earth. Flushing.
C. verticillata var. evoluta Fr., on earth. Flushing.
C. squamosa Hoffm., on earth. Flushing.
*Graphis scripta var. gracilens Nyl. Cold Spring Harbor.

Boys' High School, Brooklyn, N. Y.

*Determined by Mrs. Carolyn W. Harris.

NOTES ON VERMONT BRYOPHYTES.

A. J. Grout.

Since Mr. C. D. Howe published his list of Vermont Hepaticæ in 1899,
several other workers have taken up the study of the hepatics of our state.
Of these Dr. Alexander W. Evans has done the most work, and his notes in

Rhodora for June, August, and September, 1904, in addition to his list of New England Hepaticæ in the same journal for June, 1903, bring our knowledge of New England Hepaticæ up to date.

During the past summer I have collected and studied both hepatics and mosses and have found a few interesting things. On August 30th of last year I spent a day in Downers Glen, Manchester, a spot which I had previously visited in the days before I knew a moss from an hepatic. The wild beauty of the glen made such an impression on me then that I have always felt that a careful exploration would find it rich in rare mosses. The following list will show how well founded my belief was, and I am sure that in one short day I could only have begun to find the good things that grew there.

Mosses not Before Recorded.

Dicranella heteromalla sericea Schimp. Growing among crystals of copperas near old mine. Shrewsbury, Clifton D. Howe. A peculiar moss from a peculiar habitat. Determined by Dr. Best, but with a little uncertainty expressed. ·

Oreoweisia serrulata (Funck.) DeNot. Moist ledges, Downers Glen, A. J. G.

Rhabdoweisia denticulata (Brid.) B. & S. Moist rocks, Downers Glen, A. J. G.

Rhacomitrium canescens (Timm.) Brid. This plant covers square rods of sandy soil on the "Island" in the Connecticut River opposite Brattleboro, but all the plants seemed sterile. Although no plants of this species were found in Vermont it is more than probable that they will be found on the Brattleboro side of the river. It would seem that this plant is not so much of a subalpine as I had previously thought.

Timmia megapolitana Hedw., Downers Glen, A. J. G.; Tunbridge, Miss Mary F. Miller; Silver Lake, Brandon, Miss Annie Lorenz. It seems rather peculiar that this species not before reported from the state should be found by three different people in so widely separated localities in the same season. Is it possible that conditions for fruiting had been unusually favorable? The sterile plants would probably be overlooked as they would be mistaken for some member of the Hair-cap Family.

Thuidium pygmaeum B. & S. On face of boulder, Newfane, A. J. G. Determined by Dr. Best. ،

Hypnum montanum Wils. In large sheets covering several square feet over ledges in bed of rivulet. Downers Glen, A. J. G.

Hypnum eugyrium (B. & S.) Schimp. var. VIRIDIMONTANUM n. var. Very large, dark green throughout; leaves ovate to broadly ovate-lanceolate, about 2 mm. long. Capsules strongly constricted under the mouth when dry. Growing in wide mats over ledges in the bed of a brook in Downers Glen, A. J. G.

The relationship of this plant to *eugyrium* is clear, but seems distinct from any described variety. It may prove to be a good species. It will be issued as no. 201 of North American Musci Pleurocarpi.

Hypnum revolvens intermedium Lind. Common on the wet slides of both mountains at Willoughby, Dr. Kennedy.

Hypnum palustre laxum B. & S. On stones in brook, Newfane. A very slender form with leaves about 7 mm. long.

NEW LOCALITIES FOR RARE OR INTERESTING SPECIES.

Dicranella heteromalla (L.) Schimp. The subalpine form with curved seta, described by Mrs. Britton in the Torrey Bulletin for November, 1895, is common in Downers Glen, A. J. G.

Didymodon rubellus (Hoffm.) B. & S. Grafton, Mrs. J. D. Clapp.

Anacamptodon splachnoides (Froel.) Brid. Grafton, Mrs. Clapp.

HEPATICS NOT BEFORE RECORDED.

Pellia Neesiana (Gottsche) Limpr. A dioicous Pellia with a tubular perianth was collected by the author on soil and stones by a brook in Newfane in August. It is certainly not *epiphylla*, and Dr. Howe thinks it is *Neesiana* because of its perianth, which is shorter than in *endivæfolia*.

Fossombronia Wondraczekii (Corda) Dum. Newfane, on soil in an old road

Pallavicinia Lyellii (Hook.) S. F. Gray. Willoughby, Vt., Miss Lorenz.

Nardia obovata (Nees.) Lindb. Downers Glen, Manchester, A. J. G.

HEPATICS NOT INCLUDED IN HOWE'S LIST.

(The starred species were reported on the authority of Frost's List alone)

Riccia Sullivantii Aust. Evans List.

Riccardia multifida (L.) S. F. Gray. On moist bank by roadside, Newfane. A. J. G.

Riccardia pinguis (L.) S. F. Gray: Brattleboro, Frost. In Eaton Herbarium as *Pellia epiphylla*, Fide Dr. Evans; Jerico, Evans.

*Cephalozia divaricata (Smith) Dum. Jerico, Evans.

C. serriflora Lindb. Jerico, Evans.

*Chiloscyphus polyanthus (L.) Corda. Fide Dr. Evans.

Coleolejeunea Biddlecomiae, (Aust.) Evans. Jerico, Evans: Newfane, A. J. G.

Diplophylleia taxifolia (L.) Trevis. Mt. Mansfield, Evans; Downers Glen, Manchester, A. J. G.

Frullania Brittoniæ Evans. Jerico, Evans; Newfane, A. J. G.

F. riparia Hampe. N. Pownal, A. LeRoy Andrews.

Jungermannia lanceolata, Lake Willoughby, Lake Dunmore, Miss Lorenz: Newfane, A. J. G.

J. pumila With. Jerico, Evans.

Lophozia incisa (Schrad.) Dum. Jerico, Evans: Willoughby, Miss Lorenz.

L. inflata (Hudson) M. A Howe. Evans List.

L. Floerkii (Web. & Mohr.) Schiffn. Mt. Mansfield, Evans.

L. Marchica (Nees.) Steph. Jerico, Evans.

Marsupella sphacelata (Gieseke) Dum. Mt. Mansfield, Evans, D. C. Eaton.

Mylia anomola (Hook.) S. F. Gray. Jerico, Evans; Willoughby, Miss Lorenz.

Nardia crenulata (Smith) Lindb. Newfane, M. A. Howe, A. J. G.

N. hyalina (Lyell) Carrington. Newfane, M. A. Howe.

Scapania irrigua (Nees) Dum. Jerico, Evans.

S. paludosa (C. Muell.) Fib. Mt. Mansfield, Evans.

Sphenolobus exsectaeformis (Breidl.) Steph. Jerico, Evans.

S. exsectus (Schmid.) Steph. Willoughby, Miss Lorenz.

Anthoceros leavis L., Newfane, Burlington, A. J. G.

A. punctatus L., Newfane, A. J. G.

Notothylas orbicularis (Schwein.) Sulliv. Jerico, Evans

TWO CHANGES OF NAME.

JOHN M. HOLZINGER.

RHACOMITRIUM FLETTII. Mr. Cardot after carefully examining a part of the type of this species agrees that it is new, but he considers it should be a Grimmia, being closely related to *Grimmia torquata* Grev., *G. prolifera* C. M. & K., and *G. tortifolia* Kindb., a group of species belonging to the subgenus Rhabdogrimmia of Limpricht. I accept my friends suggestion after looking over the matter, and *Rhacomitrium Flettii* may hereafter be known as **Grimmia Flettii** (Holz.) Card. See BRYOLOGIST 7:3, 1904.

BRYUM SQUARROSUM Kindb. in Hedwigia, p. 66,1896. This name is not tenable being a synonym for *Paludella squarrosa*. I propose for this Kindbergian plant **Bryum Baileyi** (Synonym *Bryum squarrosum* Kindb., non Linn.) in recognition of my very efficient bryological friend, Dr. John W. Bailey, of Seattle, Wash., who collected this plant at Lake Union near Seattle, in quantity for my Acrocarpi. I am indebted to Mr. Jules Cardot for its final determination. The nearest relative of this West American species is *Bryum Donianum* of the section Eubryum. But this has a thickened margin and smaller leaf cells. *Bryum Baileyi* has the margin of *one* cell layer, its leaf cells in the lower part of the leaf measure 24 by 120μ. The spores are smooth and measure 10 by 14μ, while Limpricht records the size of spores of *B. Donianum* as 9-10μ.

CURRENT LITERATURE.
Musci Norvegiæ Borealis by Dr. I. Hagen.
JOHN M. HOLZINGER.

The third and concluding part of this work on the mosses of Northern Norway, covering pages 241 to 382 and accompanied by twenty-four pages of Introduction appeared in February, 1905. The two plates, figure distinctive features of *Gyroweisia tenuis compacta, Distichium Hagenii, Encalypta mutica, Timmia Norvegica, T. elegans, Polytrichum inconstans* and *Hypnum fastigiatum mitodes.*

In his Vorbemerkungen (The Introduction), the author gives an interesting history of the botanical explorations of the far northerly region covered by his work with special reference to the bryological journeys in the nineteenth century, Martin Vahl and Bishop Gunnerus being the only collectors dating back into the preceding century. Even up to 1880 comparatively little had been done in giving a comprehensive view of the mossflora of this region, practically all lying in the arctic circle. Wahlenberg's "Flora Lapponica" appearing in 1812, mentions one hundred and forty seven species of Norwegian mosses. Sommerfelt's "Supplementum Floræ Lapponicæ," 1826, adds forty-two species to Wahlenberg's list. Then up to 1870 there was little active exploration, when interest was rekindled by the journeys of P. G. Lorentz, J. E. Zetterstedt, Berggren, A. Blytt, and Arnell and Jörgensen. Among the other explorers of the region up to the present time the author, mentions Añgstroem, C. Hartman, Huebener, Parry, J. W. Zetterstett—names which will ever be interesting to students of arctic mosses.

In the present work the author has set for himself the task of reporting on the collections made principally by Arnell, Fridtz, Kaalaas, Kaurin, Ryan and himself in the years 1886-1897. The six or seven years spent on these voluminous collections has resulted in the reporting of six hundred and fifty nine species and varieties, comprising four hundred and thirty-two Acrocarpi, one hundred and seventy-nine Pleurocarpi, thirteen species of Andreæa and thirty-five of Sphagnum. In this large aggregate of species the author points out that the pygmies of the moss world, the minute ephemeral species like those of Archidium, Acaulon, etc., are almost entirely missing, showing the incapacity of these minute plants to struggle against the odds of an arctic climate.

This report includes many critical notes on the more difficult genera, noteworthy among which is Orthotrichum, and especially Bryum. In the latter genus he has enumerated no fewer than one hundred and five species and varieties, three of which are new. In all, fifty-one new species and varieties are described. From Part III we mention *Timmia elegans* Hag., *Polytrichum inconstans* Hag., *Orthothecium strictum* Lor., *Brachythecium saltense* Hag., *Amblystegium versirete* Hag., *Hypnum stragulum*, Hag., and *Hypnum fastigiatum mitodes* Hag. Seven North American species were found to reach over into this arctic region of Norway, namely: *Orthotrichum Groenlandicum*, *Bryum foveolatum*, *B. polare*, *B. crispulum*, *Pogonatum dentatum*, *Pseudoleskea denudata Holzingeri*, *Hypnum Berggrenii*. Three had been known only from Spitzbergen, viz: *Trichodon oblongus*, *Bryum globosum*, *B. teres*. Fifteen other species new to this territory were previously known from other parts of Europe.

Both by reason of the critical spirit of the work and of the position of the area covered in the report, the author has done a valuable service not only to Scandinavian Bryology in the narrower sense, but in a wider sense to the Arctic Bryology of North America and Northern Asia.

For Review of Parts I and II see BRYOLOGIST, 5:1902. pp. 44. 45.

Winona, Minn.

SULLIVANT MOSS CHAPTER NOTES.

The following names have been added to the list of Chapter Members since March 1st, making the total number one hundred and forty-seven: Mr. D. M. Andrews, Box 86, Boulder, Colorado. Mrs. Henry T. Gregory, Southern Pines. N. C. Mr. J. B. Flett, 221 N. Tacoma Ave., Tacoma, Wash. Mr. E. Bethel, 270 S. Marion St., Denver, Colorado.

NOTE.

Mr. G. K. Merrill, 564 Main street, Rockland, Me., will be glad to determine any Cladoniæa sent him by Chapter Members. The specimens must be ample and accompanied by full data.

OFFERINGS.

(To Chapter Members only. For postage.)

Miss Harriet Wheeler, Chatham, Columbia Co., N. Y. *Hypnum uncinatum* Hedw., c.fr. Collected in Adirondack Mts., N. Y. *Brachythecium Starkei* (Brid.) B. & S., c.fr. Collected in White Mts., N. H.

Mrs. Agustus P. Taylor, Thomasville, Georgia. *Pagonatum brachyphyllum* Beauv., c.fr. Collected Thomasville.

Mr. N. L. T. Nelson, 3968 Laclede Ave., St. Louis, Mo. *Dicranella varia* (Hedw.) Sch., c.fr. Collected in Illinois. *Orthotrichum Porteri* Aust., c.fr. Collected in Missouri.

Mr. Ezra T. Cresson, Jr., Box 248. Philadelphia, Pa. *Pohlia Lescuriana* (Sull.) Lindb., c.fr.; *Hypnum imponens* Hedw., c.fr.; *Hypnum hispidulum* Brid., c.fr. Collected in Swarthmore, Pa.

Mr. H. Dupret, Seminary of Philosophy, Montreal, Canada. *Ceratodon purpureus* (L.) Brid., c.fr.; *Leptobryum pyriforme* Sch., c.fr. Collected in Montreal (U. S. postage accepted).

Miss Mary F. Miller, 1109 M street, N. W., Washington, D. C. *Camptothecium lutescens* B. & S. Collected in Seattle, Wash., by Dr. J. W. Bailey.

Miss Caroline C. Haynes, 16 East 36th street, New York City. *Cephalozia connivens* (Dicks.) Lindb.; *Scapania nemorosa* (L.) Dumort.

Mr. Charles C. Plitt, 1706 Hanover street, Baltimore, Md. *Porella pinnata* L. Collected near Baltimore.

Mr. Severin Rapp, Sanford, Orange Co., Florida. *Odontoschisma prostratum* Nees. Collected Sanford, Fla.

Mr. J. P. Naylor, Greencastle, Indiana. *Fissidens Julianus* (Savi.) Schimp., c.fr. Collected Greencastle.

IMPORTANT

The second edition of MOSSES WITH A HAND-LENS describes 168 species of Mosses and 51 species of Hepatics, nearly every one being illustrated.

Instead of being a book of 150 pages as at first advertised it will contain 200 pages. There are over 40 full page plates. The additional cost necessitates an increase in price and the price on all orders received after *April 1st* will be $1.75, postpaid.

NOW is the time to subscribe for MOSSES WITH A HAND-LENS AND MICROSCOPE. On and after the date of issue of the fifth part the price will become $1.25 per part. So many demands for a complete manual of the Mosses of the Northeastern U. S. that a supplement to MOSSES WITH A HAND-LENS AND MICROSCOPE will be issued immediately upon its completion. The supplement will contain full keys to all species and descriptions and illustrations of all species not included in the main work.

SEND FOR SAMPLE PAGES

A. J. GROUT ❧ 360 Lenox Road ❧ Brooklyn, N. Y.

VOLUME VIII.   NUMBER 4

JULY, 1905

THE BRYOLOGIST

AN ILLUSTRATED BIMONTHLY DEVOTED TO

NORTH AMERICAN MOSSES

HEPATICS AND LICHENS

EDITORS:

ABEL JOEL GROUT and ANNIE MORRILL SMITH

CONTENTS

Entered at the Post Office at Brooklyn, N. Y., April 2, 1900, as second class of mail
                    matter, under Act of March 3, 1879.

Published by the Editors, 78 Orange St., Brooklyn, N. Y., U. S. A.

PRESS OF MC BRIDE & STERN, 97-99 CLIFF STREET, NEW YORK

THE BRYOLOGIST

BIMONTHLY JOURNAL DEVOTED TO THE STUDY OF NORTH AMERICAN MOSSES
HEPATICS AND LICHENS.

ALSO OFFICIAL ORGAN
OF THE SULLIVANT MOSS CHAPTER OF THE AGASSIZ ASSOCIATION.

Subscription Price, $1.00 a year. 20c. a copy. Four issues 1898, 35c. Four issues 1899, 35c. Together, eight issues, 50c. Four issues 1900, 50c. Four issues 1901, 50c. Four Vols. $1.50 Six issues 1902, $1.00. Six issues 1903, $1.00. Six issues 1904, $1.00.

Short articles and notes on mosses solicited from all students of the mosses. Address manuscript to A. J. Grout, Boys' High School, Brooklyn, N. Y. Address all inquiries and subscriptions to Mrs. Annie Morrill Smith, 78 Orange Street, Brooklyn, N. Y. For advertising space address Mrs. Smith. Check, except N. Y. City, MUST contain 10 cents extra for Clearing House charges.

Copyrighted 1905, by Annie Morrill Smith.

THE SULLIVANT MOSS CHAPTER.

President, Mr. E. B. Chamberlain, Washington, D.C. Vice-President, Mrs. C. W. Harris, 125 St. Marks Avenue, Brooklyn, N. Y. Secretary, Miss Mary F. Miller, 1109 M Street, Washington, D. C. Treasurer, Mrs. Smith, 78 Orange Street, Brooklyn, N. Y.
Dues $1.10 a year, this includes a subscription to THE BRYOLOGIST.
All interested in the study of Mosses, Hepatics, and Lichens by correspondence are invited to join. Send dues direct to the Treasurer. For further information address the Secretary.

THE ONLY MOSS BOOKS

describing the mosses of the North-eastern United States that are now in print are **Grout's Mosses with a Hand-lens,** and **Mosses with a Hand-lens and Microscope.**

MOSSES WITH A HAND-LENS, Edition 2, is an octavo volume of 224 pages, printed on fine coated paper and bound in strong cloth. It describes 168 species of mosses, and 51 of hepatics. Practically every species is illustrated in the 39 full page plates and 150 other figures.

It is the only book on mosses included in the catalogue of 8,000 volumes for a Popular Library issued by the American Library Association (1904.)

To hasten sales a set of 15 common mosses will be given to the first 100 purchasers upon request, if purchase be made direct from the author at the full price and after date of this issue.

Price, $1.75. A limited signed edition of 25 copies, with actual specimens of 200 species mounted on interleaved blank pages, is in preparation. Price $15.00.

NOW is the time to subscribe for MOSSES WITH HAND-LENS AND MICROSCOPE. On and after the date of issue of the fifth part the price will become $1.25 per part. So many demands for a complete manual of the Mosses of the North-eastern U. S. that a supplement to MOSSES WITH A HAND-LENS AND MICROSCOPE will be issued immediately on its completion. The supplement will contain full keys to all species and descriptions, and illustrations of all species not included in the main work.

SEND FOR SAMPLE PAGES
A. J. GROUT ❧ 360 Lenox Road ❧ Brooklyn, N. Y.

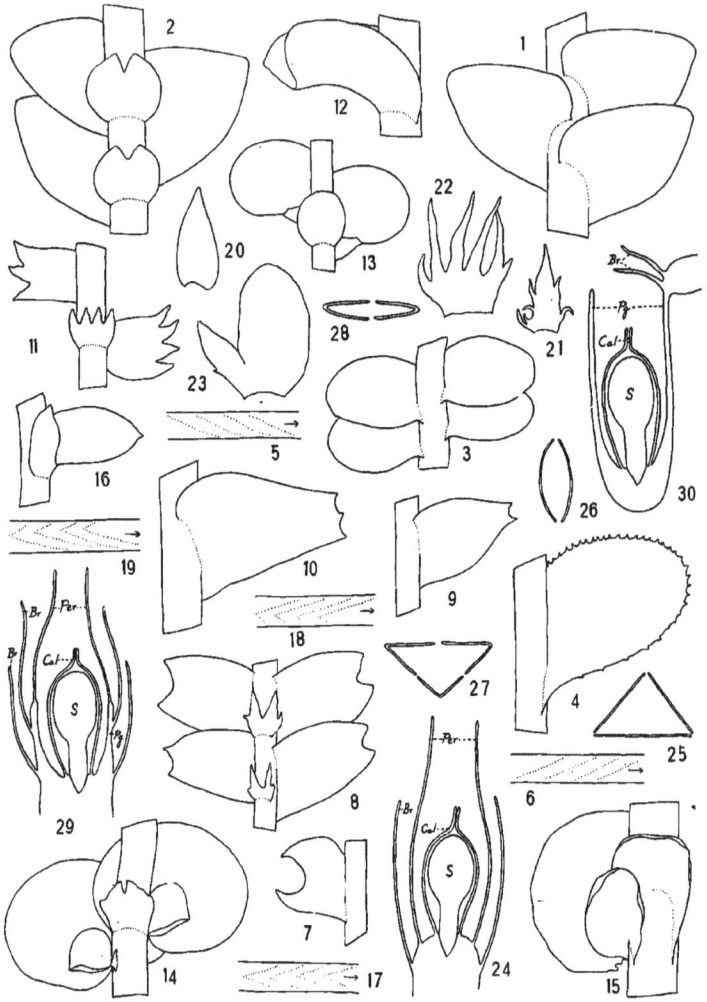

PLATE V.—*Jungermanniaceæ*.

THE BRYOLOGIST.

Vol. VIII.   July, 1905.   No. 4.

DIAGNOSTIC CHARACTERS IN THE JUNGERMANNIACEAE.

ALEXANDER W. EVANS.

[Read at the meeting of the Sullivant Moss Chapter at Philadelphia, Pa., Dec. 31, 1904.]

If one were asked to describe in a few words the general characteristics of the mosses the task would be comparatively easy, at least so far as the gametophyte is concerned. Throughout the entire group this consists of a leafy stem, and the leaves although exhibiting considerable diversity in form, in texture and in the peculiarities of the margin, are never deeply

EXPLANATION OF PLATE V.

Fig. 1. *Kantia Trichomanis*, from above. Fig. 2. The same, from below. Fig. 3. *Odontoschisma prostratum*, from above. Fig. 4. *Plagiochila asplenoides*, from above. Fig. 5. Diagram to illustrate the attachment of succubous leaves, the dotted lines representing the lines of attachment and the arrow the direction of growth. Fig. 6. Diagram to represent the attachment of incubous leaves. Fig. 7. *Cephalozia connivens*, from above. Fig. 8. *Lophocolea bidentata*, from below. Fig. 9. *Kantia Sullivantii*, from above. Fig. 10. *Bazzania trilobata*, from above. Fig. 11. *Lepidozia reptans*, from below. Fig. 12. *Marsupella emarginata*, from above. Fig. 13. *Archilejeunea clypeata*, from below. Fig. 14. *Frullania Brittoniae*, from below. Fig. 15. *Porella platyphylla*, from below. Fig. 16. *Diplophylleia apiculata*, from above. Fig. 17. Diagram to illustrate the attachment of complicate leaves which are neither incubous nor succubous. Fig. 18. Diagram to illustrate the attachment of complicate and incubous leaves. Fig. 19. Diagram to illustrate the attachment of complicate and succubous leaves. Fig. 20. *Harpanthus scutatus*, underleaf. Fig. 21. *Lophozia barbata*, underleaf (after Warnstorf); bifid underleaves also occur in this species. Fig. 22. *Cephalozia connivens*, bract. Fig. 23 *Frullania Brittoniae*, bract. Fig 24. Diagram representing a radial section through an archegonial branch and young sporophyte in the genus *Lophocolea*: *S.* sporophyte; *Cal.* calyptra; *Per.* perianth; *Br.* bract. Fig. 25. Diagram representing a cross-section of the perianth in the same genus, *i. e.*, an epigonianthous perianth. Fig. 26. Diagram representing a cross-section of the perianth in the genus *Plagiochila*. Fig. 27. Diagram representing a cross-section of the perianth in the genus *Cephalozia*, *i. e.* a hypogonianthous perianth. Fig. 28. Diagram representing a cross-section of the perianth in the genus *Scapania*. Fig. 29. Diagram representing a radial section through an archegonial branch and young sporophyte in the genus *Marsupella*: *Pg.* perigynium; other references as in Fig. 24. Fig. 30. Diagram representing a radial section through the pendent perigynium and young sporophyte in the genus *Kantia*.

The May BRYOLOGIST was issued May 1st, 1905.

lobed or cleft and never show very marked differences in the various parts of the plant, even when the stem and its branches are prostrate and closely adherent to the substratum.

With the Hepaticae, on the other hand, the task would be much more difficult. The gametophyte here exhibits the greatest variety in different families. It is sometimes a flat thallus without any indication of leaves, sometimes a thallus-like stem with rudimentary leaves, sometimes a more or less cylindrical stem with distinct leaves; and these various types are connected by intermediate conditions. In the thallose forms the cell-structure is sometimes uniform throughout or nearly so, and sometimes shows a high degree of differentiation; here again there are intermediate conditions. Both thallose and leafy species are almost always prostrate and show marked differences between the upper and lower portions. In other words they are "dorsi-ventral." In the thallose forms the dorsiventrality manifests itself in differences in cell-structure: in the leafy forms in differences between the leaves.

In the eastern United States nearly three fourths of our Hepaticae are leafy and belong to the family Jungermanniaceae, sometimes spoken of as the "scale-mosses," and we will confine our attention to these. The scale-mosses may usually be distinguished at a glance from the true mosses by the prostrate habit to which allusion has just been made and by the fact that this habit usually brings about a distinctly flattened appearance for the whole plant, the leaves themselves as well as the stem being more or less appressed to the substratum. When we examine a plant carefully we find that the leaves are more regularly arranged than in most of the mosses; looking at a stem from above (*Fig. 1*) we see two distinct longitudinal ranks of leaves spreading out on either side: looking at the same stem from below (*Fig. 2*) we can usually see a third rank of leaves, which are more or less appressed to the stem. The leaves in fact are in a spiral and conform to the one third arrangement. It will be seen at a glance that the leaves are not all exposed to the same external conditions. Those, for example, in the two ranks which we saw from above, are turned toward the light, and are, therefore, well placed for carrying on photosynthesis; those in the third rank, however, are practically cut off from the light. Probably this difference in environment has been a potent factor in bringing about a diversity in the leaves, those in the third rank being different in form and usually much smaller than the others. For the sake of convenience in description, the leaves of the third rank are spoken of as "underleaves" while those of the other two ranks are called "side-leaves" or simply "leaves." In certain of our species the underleaves are so small that they can be demonstrated only by careful dissection; in a few species they are absent altogether. Even in the last case, however, the leaves are closer together on the upper surface of the stem than on the lower, so that they do not conform to the one half arrangement, which we would naturally expect with only two longitudinal ranks of leaves.

The leaves exhibit a much greater diversity of form than the underleaves, and this manifests itself in pecularities of the margin, in lobing or division, in folding, and in the development of remarkable structures known as water-sacs. The simplest type of leaf is that which is undivided, although this is probably not the most primitive type. In this case the leaf varies in form from ovate to broadly rotund, and in all our northern genera is rounded or bluntly pointed at the apex (*Figs. 1-3*). The margin is commonly entire but is more or less toothed in certain species of *Plagiochila* (*Fig. 4*). The leaves here and throughout the group are sessile just as in the mosses, but the line of attachment instead of being transverse is usually oblique; sometimes the forward or apical end of this line is turned toward the substratum and sometimes away from it, these conditions being best shown by such diagrams as *Figs. 5* and *6*, the arrows indicating the direction of growth. These differences in the attachment of the leaves bring about differences in the way in which they overlap each other and are of the utmost importance in distinguishing certain genera. The condition seen in *Fig. 5* is known as succubous and is found in the common genera *Jungermannia, Nardia, Chiloscyphus, Plagiochila* (*Fig. 4*) and *Odontoschisma* (*Fig. 3*). The other condition is called incubous (*Fig. 6*), and is found in the common *Kantia Trichomanis* (*Figs. 1, 2*). The distinction between succubous and incubous leaves applies not only to species with undivided leaves but also to many of those with variously lobed or divided leaves.

Among lobed or divided leaves the simplest condition is found where only two apical teeth or lobes are present; sometimes the teeth are very minute and only one or two cells long; in other cases the divisions extend to the middle of the leaf or beyond. Among species with succubous leaves the bidentate or bilobed condition is found in the genus *Cephalozia* (*Fig. 7*), and in many species of *Lophocolea* (*Fig. 8*) and *Lophozia*; it is much rarer in species with incubous leaves but is clearly shown by *Kantia Sullivantii* (*Fig. 9*). In *Lophocolea heterophylla* the leaves show all gradations between the deeply bilobed and undivided conditions.

Tridentate and trifid leaves, quadridentate and quadrifid leaves are also found among the Hepaticae; none of our northern species, however, show a larger number of primary lobes than four. In the genus *Bazzania* the leaves are incubous, and in our commonest species, *B. trilobata* (*Fig. 10*), have three apical teeth. In *Lepidozia reptans* (*Fig. 11*), also with incubous leaves, the same stem will often produce both trifid and quadrifid leaves; the same is true of the succubous leaved *Lophozia barbata* and of other species of this genus. Here again gradations between bifid leaves and those just considered are also to be observed.

All of the leaves which we have so far noted are more or less flattened in one plane. The form of the leaves, however, is much more difficult to understand when the lobing is accompanied by folding. This condition is described as "complicate," the fold being called the "keel." We find it most frequently among bilobed leaves, which are then described as "complicate-bilobed." In these leaves the method of attachment is entirely different from what we have described above, each lobe being attached independently to the stem and the two lines of attachment meeting at an angle, which is sometimes very sharp. In the genus *Marsupella* (*Fig. 12*), and in certain species of *Scapania* and *Sphenolobus*, the lobes are approximately equal in size, and the leaves ought not to be described as either incubous or succubous. In the majority of cases, however, where the complicate condition occurs, the lobes are unequal in size, the dorsal lobe being the larger in certain species and the ventral in others. In the Lejeuneae (*Fig. 13*), in *Frullania* (*Fig. 14*), *Radula* and *Porella* (*Fig. 15*), the dorsal lobe is the larger and the leaves are described as incubous : in many species of *Scapania* and in *Diplophylleia* (*Fig. 16*) the ventral lobe is the larger and the leaves are described as succubous. These conditions may also be best seen by diagrams (*Figs. 17-19*). In the genus *Ptilidium* the leaves are normally quadrifid and at the same time complicate, the keel occurring between the two middle divisions. In *Pt. ciliare*, which is one of the commonest and most conspicuous species in the eastern United States, the leaves are beautifully fringed on the margin; and this condition is carried to an extreme in the still more beautiful *Trichocolea tomentella*, where the leaves present the appearance of being divided almost to the base into an innumerable number of delicate hairs.

The leaves in many of the Hepaticae, through their arrangement, overlapping, lobing and folding, doubtless assist the plant materially in absorbing and retaining water. This is seen especially well in the two species just described, where the whole plant is practically permeated in all directions by capillary spaces, which can take up and hold water like a sponge. In certain genera this function is assumed by a definite part of the leaf, which becomes hollowed out and is know as the "water-sac." Among our northern genera this structure is best studied in *Radula*, in the Lejeuneae (*Fig. 13*), and in *Frullania* (*Fig. 14*). all of which, as noted above, are characterized by complicate-bilobed leaves, the dorsal lobe being the larger. In all these forms the water-sac is formed wholly or in part by the ventral lobe or. as it is often called, the "lobule," to distinguish it from the dorsal division of the leaf, called simply the "lobe." In *Radula* and in the Lejeuneae, the free edge of the lobule is closely appressed to the lobe except in the outer part, and the region of the leaf near the keel becomes inflated and acts as the sac, the water gaining entrance through the minute opening in the outer part where lobe and lobule are not in contact. In these cases both lobe and lobule enter into the formation of the water-sac. In *Frullania* a part of the lobule itself becomes hollowed out into a hood-like organ, open at one end and blind at the other : here the entire sac is formed by the lobule.

In comparison with the leaves the underleaves exhibit much less variety, as has already been noted. They are almost invariably transversely attached to the stem, the line of attachment being straight or nearly so ; sometimes, however, they are decurrent, and the line of attachment becomes more or less arched. The latter condition is well seen in *Porella* (*Fig. 15*), where the decurrent base of the underleaf is sometimes longer than the free portion. Here again, as in the leaves, the simplest type is undivided, but the apex although sometimes broad and rounded as in *Archilejeunea clypeata* (*Fig. 13*), and in *Porella* (*Fig. 15*). is usually sharply pointed, the underleaf itself assuming a lanceolate or subulate form. These pointed underleaves are very well seen in *Harpanthus* (*Fig. 20*), where they attain an appreciable size, but they are also to be found in many other species. They become more complicated when their margins are irregularly toothed or ciliate, as in many species of *Lophozia* (*Fig. 21*).

Variously lobed and divided underleaves are characteristic of many genera. In *Frullania* (*Fig. 14*) and in *Lejeunea* the underleaves, which are subrotund in form, are bifid, sometimes to beyond the middle. In *Lophocolea* (*Fig. 8*), *Chiloscyphus, Geocalyx* and other genera, the underleaves are much narrower, but are also deeply bifid. In *Lepidozia* (*Fig. 11*) the underleaves are broad and deeply trifid or quadrifid in the larger species. In *Ptilidium ciliare* and in *Trichocolea* the underleaves resemble the leaves in being strongly ciliate along the margins of the divisions. In none of our northern species, however, do the underleaves produce water-sacs, although this phenomenon occurs among certain antarctic genera.

It will be seen from what has been said so far that the leaves and underleaves afford generic characters of much importance. These are supplemented, in the purely vegetative part of the plant, by characters derived from the branching and from the cell-structure; but, as it is difficult to discuss these without entering into considerable detail, we may pass at once to characters connected with the reproduction.

The antheridia and archegonia of the scale-mosses are essentially like those of the true mosses. The archegonia are borne singly or in groups on the tips of specialized branches, the growth of which is thereby terminated. As a rule only one archegonium of a cluster develops a sporophyte. Even in the absence of fertilization, the archegonial branches yield important characters. Instead of being prostrate such branches tend to be ascending or erect, and their leaves are oftentimes very different in appearance from the ordinary vegetative leaves. These leaves are called "bracts," and the corresponding underleaves "bracteoles." In many cases the latter are nearly or quite as large as the bracts, and this is true of some species which lack underleaves on ordinary stems. In a number of species which are destitute of underleaves, bracteoles also fail to be developed. The various species of *Radula* are striking examples of this condition. As a rule the bracts are less highly specialized or less definite in their characters than the leaves. In a species with bifid leaves, for example, the bracts tend to be irregularly two-to four-lobed (*Fig. 22*); in a species with well-developed water-sacs these structures are not developed on the bracts (*Fig. 23*); in certain species with undivided leaves, the bracts are bifid. In still other cases the bracts are scarcely to be distinguished from the leaves. Usually the archegonial branch shows a gradation between typical leaves and typical bracts.

The archegonia, the young sporophyte and the calyptra cannot be seen as a rule without careful dissection. This is because they are covered over and concealed by other parts of the plant, and are apparently thus protected from being dried up. In very rare cases, the covering is done by the bracts alone. Usually, however, in addition to the bracts, the archegonial branch bears a remarkable tubular organ called a "perianth," or else itself develops into a hollow structure known as a "perigynium."

The perianth is an organ peculiar to the scale-mosses, although not found in all of them. In the majority of our species it develops whether fertilization takes place or not; in a few species, however, its development depends upon fertilization. In no case does the development of the perianth precede that of the archegonia. The perianth consists of a hollow tube, which is attached at the base and open at the tip, sometimes with a wide mouth. Under normal circumstances it surrounds the young sporophyte and assists the calyptra in protecting it (*Fig. 24*). Most writers look upon the perianth as a structure formed by the coalescence of leaf-like organs, and it differs in appearance according to the number and character of the leaves which enter into its formation. In the simplest case it is formed by the union of two leaves and one underleaf, which remain flat. This gives rise to a perianth in the form of a triangular prism, three angles or keels being formed by the united edges of the leaves. If we suppose that such a perianth is pressed back against the substratum, two of the keels will be lateral and the third will be dorsal. This type, which is known as "epigonianthous," is beautifully shown in the genus *Lophocolea*, and may be represented in cross-section by such a diagram as *Fig. 25*. In case no underleaf takes part in the formation of the perianth, we observe a structure which is compressed at right angles to the substratum, a condition found throughout the large genus *Plagiochila* (*Fig. 26*). A very different type of perianth arises when the leaves which enter into its formation are complicate instead of being flat. In this case the keels of the leaves usually form keels on the perianth, and when a triangular perianth is formed, the third keel instead of being dorsal will be ventral. This condition, called "hypogonianthous," is found with certain modifications in a large number of genera, of which *Frullania*, *Lejeunea*, *Porella*, *Cephalozia*, *Bazzania* and *Lepidozia* may be especially mentioned (*Fig. 27*). When no underleaf takes part in this type of perianth, we again observe a compressed structure, but this time the flattening is parallel to the substratum instead of at right angles to it (*Fig. 28*). This condition is found in the genera *Radula* and *Scapania*. In many species the structure of the perianth is obscured, either by the obliteration of keels or by the interpolation of additional keels, and under these circumstances the interpretation becomes much more difficult.

The perigynium, unlike the perianth, is formed directly from the archegonial branch. It occurs in comparatively few genera, and its development is always dependent upon fertilization. While the sporophyte is growing, the archegonial branch which bears it begins to grow also in the form of a hollow tube. This encloses the young sporophyte and carries up on the outside the bracts and bracteoles. In some cases, as in *Nardia* and *Marsupella*, the perigynium bears a perianth at its mouth (*Fig. 29*). In other cases, as in *Gymnomitrium*, there is no perianth formed. In the examples so far considered the perigynium has grown in an upward direction only: in other cases, however, it grows downward as well, and sometimes its growth is entirely downward. The first of these conditions may be seen especially well in *Nardia haematosticta*, the second in *Geocalyx* and *Kantia*, the perigynium in these two genera being in the form of a pendent sac (*Fig. 30*), which penetrates into the substratum and thereby protects the sporophyte still more effectively. With this type of perigynium the perianth is almost invariably absent.

The characters noted above are usually sufficient to distinguish the genera of the Jungermanniaceae. The antheridial branches and the sporophytes occasionally yield additional characters of interest. Both of these structures, however, are likely to be uniform or nearly so throughout large groups of genera, and their characters, therefore, are more frequently tribal or even ordinal in value rather than generic. Under the circumstances it is hardly necessary to discuss them at the present time. YALE UNIVERSITY.

ADDITIONS TO THE BRYOPHYTIC FLORA OF WEST VIRGINIA.

A. LeRoy Andrews.

The "Preliminary Catalogue of the Flora of West Virginia," published by Dr. C. F. Millspaugh in 1892 (W.Va. Exp. Stat. Bull. No. 24, pp. 311-537), contained a list of eighty-three species and varieties of mosses and twenty-seven of hepatics, collected at a few points, mostly in the vicinity of Morgantown, in Monongalia County. A flora of the state embodying the results of later collections was published by Dr. Millspaugh in collaboration with Mr. L. W. Nuttall, who had made extensive collections and studies about Nuttallburg, in Fayette County (Publications Columbian Field Museum, Bot. Series, Vol. 1, pp. 65-276, 1896). In this list were noted six additional species of mosses and five of hepaticae. For both lists, as is explained in the introduction to the latter, the bryophytes had been gathered spasmodically and incidentally to the investigation of other plants.

I have seen two papers of later date listing additions to the West Virginia flora, viz.: "Some Plants of West Virginia," by E. L. Morris (Proc. Biol. Soc. Wash., Vol. XIII., pp. 171-182, 1900), and "Some New and Additional Records in Flora of West Virginia," by C L. Pollard and W. R. Maxon (Proc. Biol. Soc. Wash., Vol. XIV., pp. 161-163, 1901). Of these the former mentions two additional hepaticae and four mosses, two of which are included in the previous lists. The latter includes as new, two hepatics and seven mosses, one a repetition from the preceding paper.

From collections made mostly during the fall of 1903 and spring of 1904 in the vicinity of Morgantown I am able to add the following. Those recorded from near Masontown are from Preston County; the others, unless expressly stated, from Monongalia County. The region of Chestnut Ridge was most productive in bryophytes, the territory westward being very poor in species. Chestnut Ridge enters West Virginia from Pennsylvania, its direction slightly southwesterly, its altitude approximately 2,000 feet, and represents, so far as Pennsylvania and northern West Virginia are concerned, the extreme western ridge of the Allegheny system. East of Morgantown this ridge is cut by the valleys of Decker's Creek and the Cheat River, and the richest collecting grounds are along the mountain streams tributary to these rivers. Especially are the steams descending the western side of the ridge characterized by rapid falls and the presence in their beds of many

large sandstone bowlders, covered commonly with a luxuriant growth of
hepatics. Tibbs Run, which joins Decker's Creek at Dellslow, was explored
by Dr. Millspaugh. Quarry Run, a mountain tributary of Cheat River,
proved of no less interest, duplicating in the main the species of the other,
even to the presence of the uncommon hepatic *Herberta adunca* (Dicks.)
S. F. Gray.

The species not previously observed are the following:

MUSCI.

Sphagnum quinquefarium (Braith.) Warnst. Blister Swamp, Randolph
Co. (A. H. Moore, Sept., 1904).

Sphagnum cymbifolium Ehrh., var., *squarrosulum* Bryol. Germ.
Specimens observed by Tibbs Run as well as Dr. Millspaugh's from the same
locality preserved in the Experiment Station herbarium show strongly
marked the varietal characters. Typical forms of *S. cymbifolium* occurred
at another point in the valley of Decker's Creek, and I have a specimen collected by Mr. A. H. Moore in Blister Swamp, Randolph Co.

Andreœa rupestris Hedw. Rocks at summit of Ridge, Cheat View.

Pleuridium alternifolium Brid. Ground on Dorsey's Knob, near
Morgantown.

Dicranella rufescens (Turn.) Schp. Bank by road, valley of Decker's
Creek, near Lick Run.

Dicranella varia (Hedw.) Schp. Near road, in the vicinity of Easton.

Dicranum montanum Hedw. Decayed spot in tree, near Dellslow.

Dicranum viride (S. & L.) Lindb. From fallen tree, near Quarry Run

Grimmia apocarpa (L.) Hedw. Rocks in brook, Tibbs Run.

Rhacomitrium aciculare (L.) Brid. Tibbs Run.

Hedwigia ciliata Ehrh., var. *secunda* Schp. Rocks in Tibbs Run.
Nicely fruited specimens on boulders near Dry Run. The varietal characters
strongly marked.

Drummondia clavellata Hook. Tree, near Easton.

Ulota Hutchinsiœ (Sm.) Schp. Rocks in Tibbs Run.

Webera proligera (Lindb.) Kindb. With *Dicranella rufescens*, near
Lick Run. Not fruited.

Bryum capillare L. Quarry Run.

Bryum caespiticium L. Near Dry Run.

Mnium affine Bland. Dry Run.

Fontinalis Dalecarlica Bryol. Eur. By Tibbs Run. Not fruited.

Hookeria Sullivantii C. M. This species was collected at a few points
in Tibbs Run. No fruit was in evidence, but the leaves bore regularly at the
tips a number of clavate gemmae.

Thamnium Allegheniense (C. M.) Bryol. Eur. By Tibbs Run.

Leskea obscura Hedw. Base of tree, near Morgantown.

Anomodon obtusifolius Br. & Sch. Base of tree, by Decker's Creek, near
Morgantown.

Leptodon trichomitricn (Hedw.) Mohr. Tree near Cheat River, at Ices
Ferry. Also near Masontown:

Thuidium scitum (Beauv.) Aust. Rocks by Tibbs Run.

Climacium Americanum Brid., var, *Kindbergii* R. & C. ;" Glades " near Masontown. The swampy glades furnish an environment quite similar to that of the New England coastal region, where this form flourishes.

Pylaisia intricata (Hedw.) Byrol. Eur. Trees, near Cheat River and Decker's Creek. Probably less common here than *P. velutina*, which was noted by Dr. Millspaugh.

Brachythecium luteum (Brid.) Bryol. Eur. Ground near Quarry Run. Specimens of this species were also collected further north on Chestnut Ridge, near Mt. Pleasant, Pa.

Brachythecium rivulare (Bruch.) Byrol. Eur. Ground near Quarry Run.

Rhyncostegium serrulatum (Hedw.) Jaeg. Ground about Morgantown. This is a common species of Chestnut Ridge, near Mt. Pleasant, Pa.

Rhapidostegium cylindrocarpum (C. M.) Kindb. Decayed wood,'near Quarry Run.

Hypum reptile Rich. Ground near Tibbs Run.

Hypnum uncinatum Hedw. Quarry Run.

Hylocomium brevirostrum (Ehrh.) Bryol. Eur. Not uncommon in deep brook ravines, Tibbs and Quarry Run. Fruiting at latter. Beautifully fruiting in gorge of Cucumber Falls, at Ohiopyle, Pa.

HEPATICAE.

Cephalozia serriflora Lindb. Rotton wood, near Tibbs Run. This is possibly the same as *C. Virginiana* reported by Pollard and Maxon (Sc. A. W. Evans, in Rhodora, Vol. VI., pp. 173-174).

Frullania Brittoniae Evans. Trees near Cheat River, by Ices Ferry. Also near Masontown.

Frullania Eboracensis Gottsche. Trees by Decker's Creek, near Morgantown.

Frullania squaraosa (R. Bl. & N.) Dumort. Same locality as last.

Jungermannia lanceolata L. Rocks, in Tibbs Run.

Lejeunea cavifolia (Ehrh.) Lindb. A small form growing on rocks in Tibbs Run is referred by Dr. Evans to this species.

Lepidozia sylvatica Evans (Sc. Rhodora, Vol. VI., pp. 186-189). Ground near Tibbs Run.

Lophocolea bidentata (L.) Dumort. Rocks with mosses, by Quarry Run.

Lophozia Marchica (Nees.) Steph. Specimens from wet place by road, near Easton, are so named by Dr. Evans.

Nardia crenulata (Smith) Lindb. Springy place, near Easton.

Nardia crenuliformis (Aust.) Lindb. Few specimens from rocks in Tibbs Run.

Odontoschisma denudatum (Mart.) Dumort. Decaying stumps and logs by Tibbs Run.

Odontoschisma prostratum (Swartz.) Trevis. Rocks beside Tibbs Run. *O. Spagni* listed by Millspaugh and Nuttall is evidently referrable to one or the other of these species. (Sc. Evans on Odontoschisma, Bot. Gaz., Vol. XXXVI., pp. 321, 348).

Plagiochila Sullivantii Gottsche. Earth in vicinity of Quarry Run.

Sphenobolus Michauxii (Web.) Steph. Vertical rocks at Cheat View.

Hanover, N. H.

MUSCI ARCHIPELAGI INDICI.

(PREPARED AND DISTRIBUTED BY MAX FLEISCHER.)

The seventh series of these very interesting mosses have been recently received, including numbers 300-350. They are accompanied by a printed index and each label bears the date of issue as well as of collection! They include mosses from Java, West Java and Ceylon, with a few from Borneo, Malacca and Singapore. The specimens are abundant and well prepared, the labels models of typography. Of one rare species, *Ephemeropsis Tjibodensis*, large leaves, covered with this species have been distributed. Many of the genera are familiar, but the species are almost all different. There are some new species and many new combinations in these exsiccatæ.

New York Botanical Garden.   E. G. BRITTON.

A CORRECTION.

Hypnum eugyrium var. *viridimontanum*, published in the May BRYOLOGIST, appears to be *Raphidostegium Marylandicum* (C. M.) J. & S. This was discovered some time before THE BRYOLOGIST was printed, but through a misunderstanding was not corrected.   A. J. G.

LICHENOLOGY FOR BEGINNERS--II.

FREDERICK LEROY SARGENT.

(Begun in May 1905, issue.)

Among the first specimens a student is likely to collect there will almost surely be found examples of the species known as *Parmelia conspersa*, which grows most plentifully on stone walls and rocks in pasture land. Its general form is shown in Fig 1. The upper surface is pale greenish or straw color, becoming darkened with age; the under surface is dark brown or black. Upon the upper side there are almost always to be found a number of chestnut-colored saucer-shaped fruits. This *Parmelia* will answer as a typical example from which we may gain a good idea of the essential parts of a lichen and their general structure, after which we may more profitably consider the various modifications of these parts which appear in other members of the group.

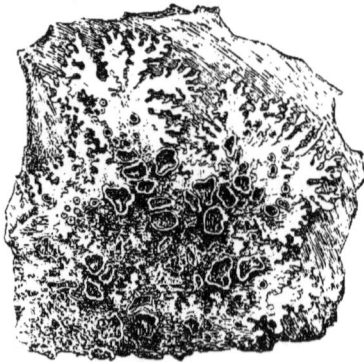

Fig. 1.
Parmelia conspersa. Natural size.
(After Rabenhorst.)

A cursory examination of our plant shows it to be a mat-like, much-lobed expansion, upon which are borne the conspicuous fruits. The latter are called *apothecia* (AP., Fig. 2); the main part of the plant-body is termed the *thallus* (Th., Fig. 2). The thallus increases in size by elongation and repeated branching of the lobes near their tips: hence the older portions are towards the center. From the under side of the thallus are developed numerous projections, called *rhizoids* (Rz., Fig. 2), which serve to attach the plant to the "substrate," or surface upon which it rests. Occasionally, upon the upper surface of the thallus appear little granular or powdery heaps (Sd., Fig. 2), called *soredia*. These are sometimes so numerous as to alter considerably the appearance of the lichen. Finally, close scrutiny with the magnifier will bring to view a number of black specks scattered irregularly over the lobes. Each is the mouth of a small cavity, called a *spermagone*, which extends into the body of the thallus (Sg., Fig. 2).

If a very thin slice cut across the thallus be examined with a magnifying power of two or three hundred diameters, the structures shown in Fig. 3 are

Fig. 2.

The same. Digrammatic vertical section. Th., thallus; Rz., rhizolds: Sd., sorendia; Ap., apothecium; Sg., spermagone. (Original.)

exhibited. The principal mass is composed of delicate tubular threads, called *hyphæ* (H), which are rather loosely interwoven to form an inner or *medullary layer* (M), and firmly compacted toward the surface of the thallus, forming a *cortex* (C) above and below. Just below the upper cortex is an irregular layer composed of innumerable bright green bodies (G) interspersed among the medullary hyphæ. These little bodies are termed *gonidia*. In each we may distinguish a transparent envelope surrounding the green protoplasmic contents. Close examination will show that branches of the medullary filaments are often in intimate contact with the gonidia.

For many years after their discovery it was believed that this close connection indicated that the gonidia were outgrowths from the hyphæ or *vice versa*. In 1869, however, the great German botanist Schwendener showed there were strong reasons for believing that the gonidia are not genetically connected with the hyphæ, but are minute *Algae*, upon and around which had grown the hyphæ of a parasitic fungus; in other words, that a lichen is not a single individual possessing as organs hyphæ and gonidia, but is a community consisting of (1) a host of small *Algae*—such as one finds growing by themselves on trees and rocks—and (2) a fungus, the like of which is also found living separately upon bark, but which in a lichen has become adapted to imprison *Algae* and gain nutriment from them.

Ever since Schwendener's time important evidence has been accumulating to confirm his view, until to-day it appears to be as well proved as Harvey's theory of the circulation of the blood. Without going at length into the details of this evidence, we may cite, in brief, the following facts:

1. All known forms of gonidia have been found to resemble species of *Algae* (belonging to several diverse families) which grow in situation

—68—

favorable to their being attacked by lichen-fungi. The only differences between the gonidia and the free *Algae* are such as would naturally follow from their different conditions of life.

2. What the theory considers to be the fungal part of lichens agrees in most important particulars with certain non-lichenous fungi, belonging to three different orders; and while the fungal part of some lichens differs considerably from the other fungi of their order, there are all gradations between these species, and some which cannot be distinguished from non-lichenous forms, except by their growing in contact with gonidia.

3. Gonidia have been separated from lichens and made to grow by themselves, when they exhibit all the characteristics of free *Algae*.

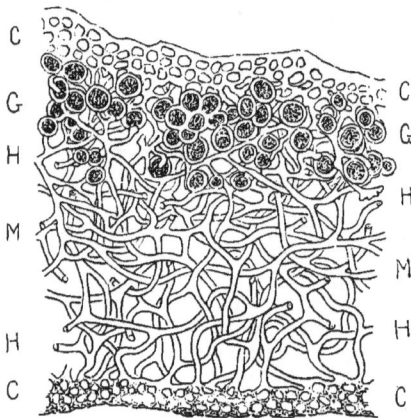

Fig. 3.

The same. Vertical section of thallus, magnified about 250 diameters. H., hyphæ; M., medulla; C. cortex; G., gonidia. (Original)

4. Lichen-fungi have been made to grow without gonidia, like other fungi, by supplying them with organic food in solution.

5. Lichen-fungi have been made to grow upon free *Algae*, and upon gonidia taken from other species of lichen, and produced a regular lichen-thallus.

6. The hyphal part of certain lichens is for some time entirely without gonidia, and gains its nutriment from bark, like other fungi; later it feeds on *Algae*.

Regarding the function of gonidia, there is but one opinion, namely, that they are the food producers of the little community, and give of their abundance to the hyphæ, which latter in the absence of some such supply of organic material could not live. We may compare the lichen-fungus to a farmer, and the gonidia to his cattle which yield him food, while he in turn shelters them and otherwise provides for their necessities.

There can be no doubt that the gonidia thrive under the conditions imposed upon them, for they multipy so rapidly as to burst through the cortex, thus giving rise to the soredia before mentioned. Each soredium is, in fact, a tiny cluster of gonidia surrounded by hyphæ. When detached and carried by the wind to some moist surface favorably situated, it grows into a lichen the same as that from which it was derived. Soredia are thus little colonies sent out by a parent community. With certain species, especially in certain localities, this is the chief—if not the only—method of reproduction. It will readily be seen how admirably adapted is this method for securing wide distribution of sorediiferous lichens. Even when no *Algae* are present, as on freshly exposed rock surfaces, soredia may establish themselves.

The apothecia, or proper fruit of our typical lichen, consists essentially (see Fig. 4) of (1) a number of short hyphal branches perpendicular to the surface, and compacted into a dense layer, called the *hymenium* (Hm.), which arises from (2) a denser tangle of hyphæ, termed the *hypothecium* (Hc.). The surface of the hymenium is called the *disk* (Dk.). These parts (in Parmelia) are more or less enveloped by a continuation of the thallus (compare Fig. 2), called the *thalline exciple* (Th. Ex.).

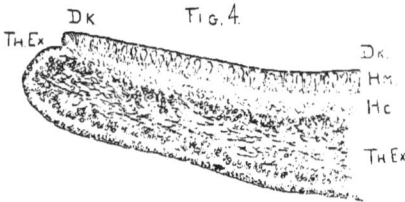

Fig. 4.
The same. Vertical section of apothecium, magnified about 50 diameters. TH., Ex., thalline exciple: Hc., hypothecium; HM., hymenium: DK disk.

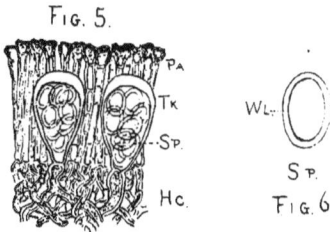

Fig. 5.

Fig. 5.
The same. A portion of the hymenium, magnified about 350 diameters. PA., paraphyses; TK., theka: SP., spore; Hc., hypothecium.

Fig. 6.
The same. A spore, magnified about 1,000 diameters. WL., wall. (Original.)

The hyphal branches which compose the hymenium are of two sorts: (1) slender filaments (*paraphyses*, Pa., Fig. 5), each ending at the surface in a colored knob, and (2) club-shaped sacs (*thekes* or *thecæ*, Tk) each of which contains when mature usually eight minute bodies known as *spores* (Sp.) The spores are reproductive bodies, each capable under suitable conditions of growing into a lichen-fungus like the parent. In our Parmelia the spores (Fig. 6) are simple cells, ellipsoid in form, and consist of a delicate transparent wall (Wl.), enclosing gelatinous (protoplasmic) contents which are colorless or but faintly tinged. The paraphyses besides affording some protection to the young thekes during development, aid in the ejection of the spores. Under the influence of moisture the paraphyses swell, and thus press upon the ripe thekes so that the apices are ruptured and the spores squeezed out with considerable force. A melon seed pressed between thumb and finger illustrates well what happens.

Cambridge, Mass.

(To be Continued.)

TORTULA PAGORUM (MILDE) DeNOT.

WILLIAM EDWARD NICHOLSON.

I was interested in the record in the July, 1904, number of THE BRYOL-OGIST by Dr. Grout of *Tortula pagorum* from Georgia, and, as shortly after receiving it, in company with Mr. H. N. Dixon, visited Milde's original locality for this species I have thought that a few notes on its occurrence there might not be without interest. I am a little surprised that Dr. Grout should predict a wide and more northern distribution, for this species, since in Europe it has a very limited and southern distribution, being apparently confined to Milde's original locality at Meran in the South Tyrol, where it is found in the same district as *Timmiella anomala* DeNot., *Fabronia octoblepharis* Schwgr., *Thuidium pulchellum* DeNot., *Braunia sciuroides* Bry. Eur. and other southern mosses. It is mostly found on dry exposed rocks and only occasionally on trees. We noticed it principally on a dry vineyard wall in the neighborhood of Algund, a village close to Meran, where it presented a very dusty and dried-up appearance.

In all probability *T. pagorum* is only a specialized form of *T. lævipila* DeNot. adapted to xerophytic conditions and it is connected with the type by the subspecies or variety *T. lævipilæformis* DeNot. This form has a wide distribution in Europe and extends to England. I find it not infrequently in my own neighborhood. Several characters have been predicated of this form as separating it from typical *T. lævipila*, but these seem to be rather unstable, with the exception of that derived from the presence of brood-leaves in the center of the terminal rosettes. These brood-leaves are well described by Prof. Correns in his " Untersuchungen über die Vermehrung der Laubmoose durch Brutorgane und Stecklinge" (p. 85 et seq.) and are most interesting, as they throw light on the origin of the somewhat similar bodies in *T. pagorum*. In *T. lævipilæformis* they are obviously modified leaves. Intermediate stages occur between them and the true leaves, and in fact they never entirely lose their leafy character. In *T. pagorum* the specialization has proceeded much further; their appearance is very different from that of the true leaves and they are much more uniform.

It is probable that *T. lævipilæformis* will also be found in North America, as it probably follows the greater portion of the distribution of *T. laevipila*, though it seems to be rather more frequent in the southern portions of the area inhabited by that species.

In the first volume of his work (Laubmoose, &c., Vol. I. p. 683) Limpricht hazards the suggestion that *T. pagorum* may be a forma *propagulifera* of *T. alpina* Bruch., which is common at Meran, but our observations on the spot did not at all give color to this view, which Limpricht apparently withdrew subsequently, since in his supplemental note on *T. lævipilæ. formis* (Laubmoose, &c., Vol. III, p. 707) he expressly compares the brood-leaves of this species with those of *T. pagorum*.

Lewes, Sussex, England.

A LONG LOST GENUS TO THE UNITED STATES—ERPODIUM (BRID.) M. C.

By a strange series of accidents and mishaps a rare moss which was collected by W. S. Sullivant in Georgia sixty years ago, and was described by Austin thirty-two years later, has remained in oblivion ever since 1877 and has only been rediscovered in connection with my studies of West Indian mosses! Owing to its resemblance to *Frullania* or *Lejeunia*, it had been sent to Manchester, England, with Austin's Hepatics which were sold to W. H. Pearson. Subsequently it was returned to the Herbarium of Columbia University and placed among the Hepaticæ, where Dr. Howe rediscovered it. Dr. Evans has supplied me with the following references, and the description is drawn from Austin's specimens.

Erpodium biseriatum (Austin) Austin Bot. Gaz. **2**:142, 1877.

Lejeunia biseriata Austin Proc. Acad. Sci. Phila. **21**:225. 1869.

Stem slender, 1 cm. long and about 1 mm. wide. Leaves 0.40–0.50 mm. long, unequal at base, with distinct hexagonal or rounded cells at apex. 0.005 x 0.013 mm. in diameter with thick brown walls, basal and central cells longer and narrower, 0.010 x 0.040 mm., the translucent marginal cells not papillose, dorsal cells with from 4–8 minute papillæ. Fruit unknown.

Collected with *Lejeunia Sullivantii* by W. S. Sullivant, near Augusta, Georgia, in 1845.

Dr. Small tells me that the region around Augusta is very hot and moist, with densely wooded river swamps, where mosses and hepatics abound. This would account for the occurrence of this tropical *genus* within our limits, as its nearest relative *E. Cubense* and *E. Domingense* are in Cuba, Santo Domingo, Porto Rico and Jamaica, with another species, *E. diversifolium*, in Mexico. Full descriptions of these will be found in the Bulletin of the Torrey Botanical Club for May. ELIZABETH G. BRITTON.

NEW OR UNRECORDED MOSSES OF NORTH AMERICA.

By J. CARDOT AND I. THÉRIOT.

Translated and condensed from The Botanical Gazette, May, 1904.

Descriptions of new species given in full. See BRYOLOGIST, January and March, 1905.

BARTRAMIA ITHYPHYLLA Brid. var. FRAGILIFOLIA Card. & Thér.

Differs from the type in its rigid, fragile, much broken leaves.

Colorado: Along the Cogwheel Railroad to Pike's Peak, alt. 2100–3000m. (J. M. Holzinger, 1896).

By its brittle and usually broken leaves, this form much resembles *B. breviseta* Lindb., but in the latter the leaf base is hardly glossy and less abruptly contracted to the subula.

WEBERA CHLOROCARPA Card. & Thér.

Rather densely caespitose, covered with soil at the base, fuscous green below, above yellowish. Stems 1-2 cm. long, erect, simple or divided.

Leaves erect-appressed, 2 mm. by 1 mm., ovate-lanceolate, acutely acuminate, a little decurrent at base, margins plane and entire, costa 80μ thick below, short excurrent, basal cells quadrate or short-rectangular, subinflated, 40-60μ by 25-40μ, median hexagono-rhomboidal 40 by 18-20μ, four to five rows, marginal cells of the upper two-thirds narrow, linear forming a sort of yellowish margin. Seta reddish, pale above, more or less flexuous, 2-2.5 cm. long: capsule 2-2.5 mm. long by 0.75 mm. wide, pendulous or cernuous, ovate-pyriform, narrowed into a neck equalling the sporangium, pale yellow, plicate with age, scarcely constricted under the mouth, walls soft with numerous superficial stomata, operculum convex, obtusely apiculate, annulus broad. Teeth 0.44 mm. high, formed of 20-25 joints, basal membrane very wide extending more than half the length of the teeth, segments widely open along the keel, cilia one or two more or less elongated, granulose. Spores 18-20μ in diameter. Seemingly dioicous (antheridial buds not seen on fruiting plants). Plate XX.

Nevada: Marlette Lake, Washoe Co., on stream bank (C. F. Baker, 1902).

Resembles in habit *W. gracilis* DeNot., but much stronger, with a very different areolation of broad and short cells. The leaf areolation recall that of the genus *Mniobryum* Limpr., but the stomata of the capsule are superficial and the annulus is quite distinct.

WEBERA DEBATI Card. & Thér.

In loose yellowish-green tufts, with the habit of Philonotis. Stems 1.5-2.5 cm. high, densely radiculose below, tomentose above, with slender erect innovations. Lower leaves rather remote, erect-spreading. upper leaves closer appressed, about 1.3 mm. long by 0.33 mm. broad, lanceolate, acute, not at all decurrent; margins plane throughout, denticulate nearly to base; costa 40μ thick at base, vanishing below the apex; median cells linear, 140-170μ, 28-30μ broad, the lower cells broader and shorter, rectangular or subhexagonal, marginal cells longer, narrowly linear. Other characters unknown. Plate XX.

North America: Alexander Co. (Herb. L. Debat. without name of collector).

This species seems closely connected with *W. annotina* Bruch., from which it is distinguished by the larger size, the habit resembling that of a small Philonotis, the tomentose stems and the leaves plane on the margins.

BRYUM PENDULUM Sch. var. NEVADENSE Card. & Thér.

Differs from the type in the more slender capsule, similar to that of the var. *angustatum* Ren., but larger; in the convex-apiculate operculum, which is not at all conic; and finally in the leaves and costa which are green, not red at base.

Nevada: King's Canon, near Carson, along stream (C. F. Baker, 1902).

BRYUM POLYCLADUM Card. & Thér.

Synoicous, in broad dense tufts, fuscous within, bright green above. Stems short, 3-5 mm. high, branches slender, erect, numerous, arising from below the perichætium. Leaves erect-appressed, crowded, the lower short,

I m. long, 0.5 m. broad, the median and upper leaves a little larger, 1.5 mm. long, 0.5–0.6 mm. broad, not decurrent at base, ovate or ovate-oblong, short acuminate, narrowly decurrent from base to apex, denticulate above, costa strong, reddish, 60–65μ thick at base, short excurrent in the middle and upper leaves, hardly percurrent in the lower; median and upper cells short-hexagonal, 30–35μ long, 12μ broad, with incrassate walls, marginal cells linear in two or three rows, lower cells larger, laxer, rectangular, 35–50μ long, 12–18μ broad. Capsule oblong, 4–4.5 mm. long, 1–2 mm. broad, nodding or perdulous, neck abruptly contracted when moist; operculum convex-apiculate. Seta elongated, flexuous, reddish, 4–6 cm. long. Annulus broad. Teeth of peristome narrow, pale, reddish above with 18–22 lamellæ, 0.35–0.4 mm. long, 50μ broad at base, basal membrane of the inner peristome adherent ⅓ the height of the teeth, segments linear, gaping along the keel; cilia very short or none. Spores minute, pale, 12μ in diameter. Plate XXI.

Nevada: Spooner, Douglas Co., in large mats on moist banks (C. F. Baker, 1902).

This moss can be placed near *B. longisetum* Bland., but it is easily distinguished from it by the numerous sterile branches arising from below the perichaetium, the smaller leaves with a shorter acumen, the peristomial teeth, which are narrower and paler, and have more numerous lamellæ, and finally the much smaller spores.

(To be Continued.)

WHAT TO NOTE IN THE MACROSCOPIC STUDY OF LICHENS.

Bruce Fink.

Introductory Statement.

Mrs. Carolyn W. Harris has, in previous volumes of the Bryologist, given amateur lichenists a series of descriptions of the more conspicuous lichen species, which will prove helpful to workers in determinations and in fixing the main features of gross morphology. It is the purpose of the present paper to state the principal features of gross morphology, including not only the foliose and fruticose lichens, but also extending the statement to the most inconspicuous crustose species as well. In so doing, we shall confine attention to such elements of structure as may readily be seen with the unaided eye or with an ordinary hand-lens.

The Thallus.

In this study, it is but natural to begin with the vegetative tract of the lichen—the thallus. The thallus may be an erect structure, rising from the substratum; a pendulous one, hanging downward from it; a conspicuous or inconspicuous flat one, closely or loosely attached to the substratum; or an inconspicuous one, largely or even wholly imbedded in the substratum. Erect and pendulous forms are commonly called fruticose thalli, and the flat or horizontal ones may be either foliose, or crustose; foliose when somewhat leaf-like, and crustose when a closely attached crust resting on or within the substratum.

GENERAL FORMS OF THALLI.

Here we may consider such characters of the three types of thalli as may be readily seen. Beginning with the foliose forms, which the student will be likely to observe first, it will be readily noticed in comparing a number of them that they are variously lobed, or that some are quite entire at the margin. In instances where the lobing is evident, the lobes may be more or less imbricated. In both lobed and unlobed forms the margin may be wavy or crenate instead of entire, and it may be ciliate or devoid of cilia.

Passing to the fruticose thalli, which are quite as likely to attract attention, one would notice first of all whether branched or unbranched, and the manner of branching. Then attention would be attracted to the surface, and one would readily observe that in some there are small outgrowths from the main axes, other than the branches. These are flat expanses in the *Cladonias*, and called squamules. In the *Stereocalons*, these outgrowths are more irregular in form, and are known as phyllocladia.

In the crustose thalli, one would note with the eye, as a rule, simply a more or less conspicuous crust spread over the substratum, or sometimes really lying wholly or partly in the substratum, and indicated at the surface often only by a change in color. These crustose thalli will be found irregular in outline or more or less plainly orbicular, and to form a continuous or more or less broken and scattered crust. In some species the tendency is toward more orbicular forms, and in others more toward irregularity in form; but in any case, the peculiarities of the surface of bark, dead wood or rock forming the substratum will determine the form of the particular thallus to a large extent.

Lichens are a late evolution, and the forms are still quite plastic. Nevertheless, the forms, sizes and colors of lichen species are quite as constant as in many undoubted autonomies, whether plant or animal. Indeed, in many lichens the morphological characters, whether gross or minute, are quite as constant as are those of most flowering plants, and it may well be doubted whether even the *Cladonias* are very much more plastic than the members of the genus *Craetegus*, including our common hawthorns.

SIZES OF THALLI.

Having disposed of the matter of forms and positions of lichen thalli, some words are in order regarding sizes. The measurements are all given in this paper in units of the metric system, and fruticose thalli of *Usnea longissima* frequently reach 1.5 metres in length, while the foliose thalli of *Gyrophora Dillenii* sometimes reach .35 of a meter in diameter. To simplify somewhat, strands of the *Usnea* five feet long have been carefully picked out of the tangled masses hanging over the branches of trees, and specimens of the *Gyrophora* and another species of the genus have been measured which surpassed one foot across the longer way of the thallus. Both fruticose and foliose thalli may vary in size from these large forms to minute ones not more than .2 mm. in height or diameter. In the crustose thalli, we most naturally think of the spread over or within the substratum, and this may vary greatly, though the spread is seldom more than 10 cm. In these and the fruticose forms, the thickness is to be taken into account. But in the descriptions, actual measurements of thickness are very seldom given, though comparative statements are often resorted to. In the descriptions of the fruticose forms the diameter of the thallus, or branches of it are often given; and here again is a considerable amount of variation found, though very much less than that of length or distance across the thalli.

THE SURFACES OF THALLI.

After noting the size and form of the thallus the observer would naturally turn to the surface and note its general character. First, in the foliose thalli, he would note whether the upper surface is comparatively smooth or wrinkled, corrugate or pustulate; whether it bears cilia or the minute growths known as isidioid branchlets, and whether it is sorediate or not. Also, now, if not before, he must notice whether the margin of the thallus is closely attached to the substratum, or more or less ascendant. Then turning to the lower surface, it will commonly be found that it is more or less covered with the attaching organs known as rhizoids. It must be noticed whether these are large or small, whether numerous or few, and whether evenly scattered or collected into rows or in groups or other forms. Then, too, the lower surface is sometimes quite smooth, except for these rhizoids, but in other instances it will be found to be variously wrinkled or pitted, or in *Gyrophoras*, bearing vertical plates which gives strength.

In the fruticose thalli, one will find the surface smooth or more or less pitted, and in some instances it is somewhat tomentose. Then, in the *Stereocaulons*, one will find the peculiar structure known as phyllocladia, and in the *Cladonias*, the squamules. The form, size, frequency of occurrence and distribution of these organs must be noted carefully. And in the *Cladonias* especially, it is necessary to note whether the cortex of the podetium is entire or more or less broken so that it becomes areolate or even disappears over some portion of the podetium. And in this same genus, careful observation with a lens is necessary to ascertain whether any part of the fruticose portion of the thallus is sorediate or not.

Finally, turning to the crustose thalli, one will find they are also smooth or variously roughened. Those that are hypophloeodal or hypolithic simply take the contour of the surface of the substratum as do also some thin and smooth forms that are in part or wholly epiphloeodal or epilithic. Others are scurfy or granular, and these are usually rather poorly developed and thin. In thicker forms we are likely to find the warty or verrucose condition, and here and there may occur minute chinks, so that the thallus is said to be rimose or chinky, or finally the chinks may become numerous and divide the thallus into minute or small several sided areas known as areoles. Such a thallus is said to be areolate.

COLORS OF THALLI.

As compared with size and form, color is usually regarded as a rather more variable and therefore less reliable taxonomic character. Yet the colors of thalli play quite an important part in determining lichens, and though often quite variable, they must be carefully noted. Colors in lichen thalli vary all the way from a white to a black, but what we will call sea-green is the most common color. This color is a greenish-gray. Some other colors are ashy, olivaceous, brown, straw-color and various intermediate conditions as brownish-black and olive-brown, etc. And the thallus is often more or less variegated, while the lower surface is frequently of a different color from the upper. Also, in the fruticose forms the basal portion is frequently of a different color than the distal portions, usually darker. The tendency of thalli, as other lichen structures, is to darken with age, and the variations in a species may usually be traced to peculiar conditions of growth, through no very definite studies of this matter have been made. Grinnell, Iowa.

(To be Continued.)

SULLIVANT MOSS CHAPTER NOTES.

The following names have been added to list of Chapter Members since May 1st, making total number one hundred and forty-nine: Mr. William L. Sherwood, 36 Washington Place, New York City. Rev. W. W. Watts, "The Manse," Young, New South Wales, Australia.

NOTE TO MEMBERS.

It has been a great regret to me that my illness and long convalescence has prevented me from determining lichens for the members of the Sullivant Moss Chapter. As I do not expect for another year to be able to do this work, Mr. Merrill has kindly consented, not only to determine Cladonias for Chapter Members, but any lichens sent to him, "providing ample specimens with full data are sent."

Mr. Merrill's address is G. K. Merrill, 564 Main street, Rockland, Maine.

The interest in the study of the Lichens seems to be increasing, and with the help given by the excellent articles published in THE BRYOLOGIST, the students ought to increase in number as well as in knowledge of these very interesting plants.

With grateful acknowledgement of the many kind messages sent me by the Chapter Members during my illness, I am, cordially,

CAROLYN W. HARRIS.

OFFERINGS.

(To Chapter Members only. For postage.)

Mr. J. W. Huntington, Amesbury, Mass. *Dicranum Bergeri*, Bland., c.fr.; *D. montanum*, Hedw., st. Collected in Amesbury. *Dicranum spurium* Hedw., st. Collected in Weare, N. H.

Mr. B. D. Gilbert, Clayville, Oneida Co., N. Y. *Camptothecium nitens* Schimp., st.; *Fissidens taxifolius* (L.) Hedw., c.fr. Collected in Clayville. (A stamp is preferred to an addressed envelope).

Mrs. Augustus P. Taylor, Thomasville, Ga. *Ditrichum pallidum* (Schreb.) Hampe., c.fr.: *Fissidens polypodioides* Hedw., c.fr. Collected in Thomasville.

Miss Caroline C. Haynes, 16 East 36th street, New York City. *Blepharostoma trichophyllum* (L.) Dumort.; *Porella platyphylla* (L.) Lindb. Collected in the southwestern Adirondack Mts., N. Y.

Mrs. R. H. Carter, 37 Church street, Laconia, N. H. *Usnea barbata* (L.) Fr. var. *rubiginea* Michx. Collected in Laconia. *Evernia vulpina* (L.) Ach. Collected in Oregon.

Mrs. Carolyn W. Harris, Chilson Lake, Essex Co., New York. *Solorina saccata* (L.) Ach. Collected Chilson Lake.

VOLUME VIII.   NUMBER 5

☀ SEPTEMBER, 1905 ☀

THE BRYOLOGIST

AN ILLUSTRATED BIMONTHLY DEVOTED TO

NORTH AMERICAN MOSSES

HEPATICS AND LICHENS

EDITORS:

ABEL JOEL GROUT and ANNIE MORRILL SMITH

CONTENTS

Entered at the Post Office at Brooklyn, N. Y., April 2, 1900, as second class of mail
matter, under Act of March 3, 1879.

Published by the Editors, 78 Orange St., Brooklyn, N. Y., U. S. A.

PRESS OF MC BRIDE & STERN, 97-99 CLIFF STREET. NEW YORK.

THE BRYOLOGIST

BIMONTHLY JOURNAL DEVOTED TO THE STUDY OF NORTH AMERICAN MOSSES
HEPATICS AND LICHENS.

ALSO OFFICIAL ORGAN
OF THE SULLIVANT MOSS CHAPTER OF THE AGASSIZ ASSOCIATION.

Subscription Price, $1.00 a year. 20c. a copy. Four issues 1898, 35c. Four issues 1899, 35c.
Together, eight issues, 50c. Four issues 1900, 50c. Four issues 1901, 50c. Four Vols. $1.50
Six issues 1902, $1.00. Six issues 1903, $1.00. Six issues 1904, $1.00.

Short articles and notes on mosses solicited from all students of the mosses. Address manu-
script to A. J. Grout, Boys' High School, Brooklyn, N. Y. Address all inquiries and sub-
scriptions to Mrs. Annie Morrill Smith, 78 Orange Street, Brooklyn, N. Y. For adver-
tising space address Mrs. Smith. Check, except N. Y. City, MUST contain 10 cents extra
for Clearing House charges.

Copyrighted 1905, by Annie Morrill Smith.

THE BRYOLOGIST.

Vol. VIII.   SEPTEMBER, 1905.   No. 5.

NOTES UPON MARYLAND BRYOPHYTES AND ON TWO MOSSES FROM VIRGINIA.

EDWARD B. CHAMBERLAIN.

During the past year there have come to me for determination several packages of bryophytes which were collected at Plummer's Island, Maryland. This island is situated in the Potomac river, about nine miles from Washington, D. C., and has been leased by the Washington Biologist's Field Club, which is now engaged upon a preliminary survey of the plant and animal life found thereon. Thus far but little systematic collecting has been made among the mosses and hepatics, and most of the specimens belong to the common species. The following, however, seem to be of sufficient interest to warrant special mention. Only one is reported in the "Guide to the Flora of Washington and Vicinity," by L. F. Ward (1881), which, as far as I know, is the last local flora including the bryophytes of this region.

Specimens of all the species mentioned are in my own herbarium and in that of Mr. E. L. Morris. Eventually, duplicates will be placed in the National Herbarium. I have to thank Dr. A. W. Evans for assistance in determining the Ricciaceae.

Aphanorrhegma serratum Sull. A few patches of this moss were found last fall by Mr. E. L. Morris and myself upon the mud-flats exposed by the low stage of the Potomac river. A small amount was collected also upon the adjacent Maryland shore, and on the Virginia shore opposite.

Ephemerum serratum (Schreb.) Hampe. A few plants of this moss were found intermingled with the *Aphanorrhegma*, at the southern end of the island. Probably the minute size of the plants renders them often overlooked, for, while it is said to be common, it is but rarely reported.

Grimmia campestris Burchell (*G. leucophaea* Grev.). Abundant upon bare ledges, but apparently always sterile. A few miles further up the river, at Great Falls, Maryland, the same species is even more abundant and occasionally fertile.

Thelia Lescurii Sulliv. On sandy ground in dry situations and abundantly fruited.

Riccia crystallina Schwein. This species, together with *R. Sullivantii* Austin, grew upon the wet bare mud of the river bank. Dr. Evans writes that *R. crystallina*, though widely distributed, is but rarely collected in the United States. Both species were in fruit.

Ricciocarpus natans (L.) Corda. The terrestrial form known as *Riccia lutescens* Schwein., grew quite commonly with the *Riccias* mentioned above, being very conspicuous because of its large green crinckled thalli, often more than an inch in diameter. As *Riccia lutescens*, it is reported in Ward's

The July BRYOLOGIST was issued July 15th, 1905.

List, mentioned above, and in the National Herbarium there are specimens collected at Rosslyn, Va., by Prof. F. V. Coville in 1889. This station is but a short distance from Plummer's Island. There is also a packet in the National Museum labelled *Riccia lutescens*. collected by Mr. Rudolph Oldberg in Rock Creek Park, but from a hasty examination of the specimens, which are scanty and broken, appear to be rather some form of *R. Sullivantii.*

NOTES ON TWO MOSSES FROM VIRGINIA.

To those interested in the ranges of North American mosses, the following stations, which have recently come to my notice, may be of value. Both mosses were collected by Mr. W. R. Maxon, in Fairfax Co., Virginia, opposite Cabin John, Maryland, about six miles above Washington. Specimens are in the National Herbarium and in my own collection.

Mnium stellare Reich. The only report of this species from this vicinity, which has come to my knowledge, is that in Ward's "Flora of Washington and Vicinity," where no data whatsoever are given. The basis of this report is probably two specimens in the National Museum, collected by Mr. Rudolph Oldberg, at "Rock Creek, near Washington," since the bryophytes of Mr. Ward's Flora were practically reprinted from a list prepared by Mr. Oldberg for the "Flora Columbiana." The moss is northern in its general range, the nearest stations of which I have record being Philadelphia and Chester, Pennsylvania, reported by Dr. Small in the "Catalogue of the Bryophyta and Pteridophyta Found in Pennsylvania." The Washington stations may represent the extreme southern range in the costal plain.

Anomodon minor (Beauv.) Fuern. This species is not listed in Ward's Flora, and I have not succeeded in finding any reports of its occurrence in this vicinity. In the National Herbarium, however, there is a specimen from Rock Creek Park, collected in 1892 by Prof. J. M. Holzinger. Lesquereux and James in the Manual remark, "in the Middle States, common." Its range seems to be much more extensive, since in my own herbarium there are specimens from Maine, Massachusetts, Rhode Island, Ohio and Minnesota, while in the National Herbarium there are also specimens from Ottawa and Ontario, Canada; Connecticut, Pennsylvania and Virginia.

Washington, D. C.

NOTES ON LUZON MOSSES.

R. S. WILLIAMS.

In walking about Manila one is rather surprised at the scarcity of mosses. The walls of the old city are well covered in places with numerous shrubs, herbs, grasses and some ferns, yet I have only observed a single species of moss on either walls or tree trunks, while the ground everywhere seems absolutely free of them. This one moss is apparently a small *Barbula* that rarely fruits.

Across Manila Bay, along the Lamao river, and up that stream to the summit of Mt. Mariveles, a region I spent some months in, a fairly good collecting ground for these plants may be found. Bushes and small trees grow from the water's edge, and back two or three miles, rather heavy forests occur, that extend with gradually diminishing size of trees to the mountain summit, yet in this apparently favorable region but very few mosses were found for the first two hundred or three hundred feet above sea level. Two species of *Fissidens*, one of large size and sterile: the other small and commonly fruiting, were rather abundant; also a *Dicranella*. A single *Bryum*, *B. coronatum*, apparently rare, was collected, also a *Neckera*, quite common on trees.

As one ascends the stream, above three hundred or four hundred feet elevation, the rocks become fairly well covered, especially with Barbulas, various Hypnoid species, and some others. A *Webera* (Diphyscium) was found on boulders, from about four hundred to one thousand feet elevation. Toward the summit of the mountain, which has a height of some four thousand five hundred feet, both trees and rocks bear numerous specimens of the true mosses, as well as liverworts. A little below the summit, on a patch of ground that had been burned over, *Funaria calvescens* was flourishing, and the combination of black and yellowish green had a most familiar appearance.

On leaving the region I went due north some one hundred and fifty miles to Baguio, with an elevation of five thousand two hundred feet. The mountain near, known as Santo Tomas, rises about three thousand feet higher. About the town are grass covered hills, alternating with open pine forests. Several species of oak are found, while near the mountain summit large species of yew and juniper flourish, as well as various other genera, well known in temperate climates. The moss flora was found to be fairly abundant, although I should estimate that scarcely one-half the number of species existed that might be found in similar regions of the north. Among other genera noted, are the following: Sphagnum, Trematodon, Ditrichum and Garckea, the latter a small tropical genus. Several Dicranella were found growing abundantly on cut banks and moist open ground, but Dicranum seemed to be rare, one or two species possibly occurring on trees. Campylopus, Barbula, Leucobryum, *Octoblepharum abidum* and Micromitrium are all well represented, either in species or individuals. Grimmia, Rhacomitrium and Orthotrichum seem to be wanting, but I have one species that looks much like an Encalypta. Bryum and Mnium are comparatively rare, both in species and individuals. *B. argenteum* occurs, also a Mnium near *rostratum*. Rhodobryum I have represented by a single species, also Catharinea, and *Rhizogonium spiniforme* is common. Along trails and on damp shady ground a Pogonatum is as common as in such situations in the United States, and Polytrichum occurs, but not so commonly. Several species of Neckera were obtained, one tree species, with stems eighteen or twenty inches long and broad, rugose, complanate leaves, being about the handsomest moss obtained. On rocks I found what looks much like *Papillaria nigrescens*, sterile as usual, and on trees were various species of Meteorium. Of the Hypneae, there is a fair proportion in the collection. A number belong to the genus Thuidium, others apparently to Plagiothecium, Hypnum, Sematophyllum, etc.

Various species that grow commonly near the summit of Mt. Mariveles also occur on the upper slopes of Mt. Santo Tomas, some three thousand or four thousand feet higher, their habits being regulated, evidently, by the more or less similar conditions of moisture, rather than by elevation.

Perhaps the most widely distributed species collected is a Fissidens, about equalling *grandifrons* in size. It is common at not much above sea level, but always sterile, while from four thousand to seven thousand five hundred feet elevation, fruiting specimens are abundant.

Manila, P. I., January 20, 1905.

BRYUM FOSTERI, n. sp.

Bryum Baileyi is not tenable (See BRYOLOGIST, 8: May, 1905). Dr. Brotherus having given this name to an Australian moss. Therefore, I propose *Bryum Fosteri* for the Washington moss: Synonym *Bryum Baileyi* Holz. non Broth.   JOHN M. HOLZINGER.

THE BOTANICAL CONGRESS AT VIENNA.

ELIZABETH G. BRITTON.

It has become a settled custom to hold an International Botanical Congress once in five years. There have been held one at Genoa, one at Paris, and the last at Vienna, from the eleventh to the eighteenth of June, at which there was an attendance of about six hundred persons, of which about four hundred were professional botanists, and nearly two hundred whose names are familiar in botanical literature. The opening exercises were held in the great hall of the university, and the morning sessions were devoted to the reading of papers, illustrated by lantern slides, and to the sessions of various societies, including the International Society of Botanists. The afternoon sessions were held at the Botanical Garden, beginning at three and ending at seven or later. They were devoted to questions of nomenclature and the discussions were based on the "*Synoptical Text*," prepared by Mr. John Briquet, who with infinite patience had brought together and coördinated the diverse views which have so confused the question of plant names. His linguistic facility won the admiration of all.

The report was presented in the name of the International Nomenclature Commission, appointed in Paris in 1900, which was printed in a quarto volume of one hundred and fifty-nine pages and contains the laws of 1867, with subsequent additions and recommendations of the International Nomenclature Commission. The official language of the session was French. M. Flahault, of Montpelier, acted as president, with two vice-presidents, Mr. Rendle, of London, and Carl Mez, and three secretaries, English, French and German. There were twenty-six German delegates, seventeen Austrian, fourteen American, eight French, eight Swiss, four Russian, three Belgian, two English and two Italian, and one each from Norway, Sweden, Spain, Denmark, Java and Calcutta. But this did not represent the total number of votes cast because a number of the delegates represented several societies and institutions, and some were reported to have as many as eleven and twelve votes each, and three to seven votes was not unusual. However, the proportions remained about the same and the preponderance of votes rested with the Germans and Austrians.

The first session was devoted to preliminaries of organizing, and it was decided to postpone consideration of all questions pertaining to fossil plants and to the mosses and thallophytes until the next Congress, and that they be referred to a special commission to report in 1910 at Brussels. Six meetings were held, all well attended, and the results reached have been characterized "*as conservative but progressive.*" The priority of the specific name was adopted but the oldest generic name met with strong opposition and a list of four hundred exceptions, with the possibility of future additions and corrections, was adopted by vote of 118 to 37. Another surprising decision, that after January, 1908, all descriptions of new species must be accompanied by a short diagnosis in Latin, was adopted by a vote of 125 to 56. Several remarkable things happened during the sessions, one of which was the first attempt to use an evident majority by putting to vote without discussion, the first fifty-two articles of the code. This met with such strong opposition that it was abandoned, and the articles were each voted on separately.

The most sensational feature was the protest by Dr. Otto Kunze against its methods, representation, votes, decisions and recommendations of the commission. This was printed in three languages, and on the fourth day Dr. Kunze appeared in person and was listened to for ten minutes, while he read his protest. When the allotted time expired he was called to order and took his leave. He characterized the methods as dishonest, and stationed men at the door of the offices of the Congress to distribute his circular. It was rather surprising to see how calmly the members accepted his criticisms and how strongly the majority felt as a reaction against his procedure. It was evident, however, that European botanists have not begun to understand the principal of generic types, nor the absurdity of an arbitrary list of exceptions.

The hope has been expressed that the Vienna Code will be followed until something better is accepted, but it seems evident that English botanists are likely to follow the Kew Rule and Kew Index, and that newer American School will not give up a definite set of principles for arbitrary exceptions.   New York Botanical Garden.

LICHENOLOGY FOR BEGINNERS—III.

FREDERICK LeROY SARGENT.

(Begun in May, 1905, issue.)

Once set free and in the presence of sufficient moisture, air, and warmth, the spores germinate by sending out one or more tubular projections (rudimentary hyphæ) which branch and elongate until the food-supply stored in the spore is exhausted. Then if they do not come in contact with *Algae*, which may serve as gonidia, they perish. But if proper *Algae* are encoun tered, then the hyphæ begin to grow vigorously and form a network of branches enveloping them. This first thin layer of hyphæ, called the *hypothallus*, remains a prominent feature of some adult lichens. With the

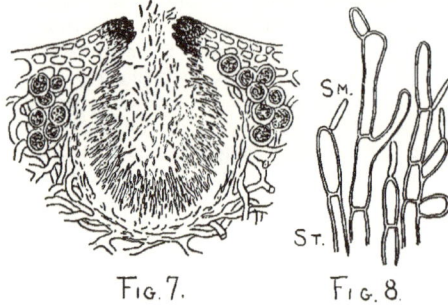

Fig. 7. The same. Vertical section of a spermagone, magnified about 200 diameters.
Fig. 8. The same. Sterigmata (ST.) and spermatia (SE.), magnified about 1,500 diameters. (Original).

majority, however, as with *Parmelia*, it serves chiefly as a groundwork from which the thallus proper is developed, although it may persist to some extent in the rhizoids.

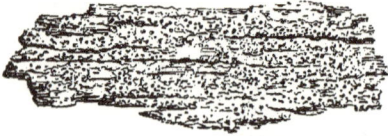

Fig. 9. *Acolium tigillare.* Slightly magnified. (Original).

The spermagones (Fig. 7. and SG., Fig. 2) are, as we have seen, flask-shaped cavities, opening by a minute pore. Into the cavity project innumerable short hyphal branches, called *sterigmata*, (ST., Fig. 8), which produce as outgrowths exceedingly minute bod. ies, termed *spermatia* (SM.). These, when ripe separate readily from the sterigmata, and under the influence of moisture are extruded through the pore in a mass of jelly. The function of the spermatia is .somewhat obscure, their minuteness rendering investigation of them particularly difficult. In certain lichens there is evidence that they perform a service analagous to that of pollen—that is to say, they are male elements that fertilize a female cell,

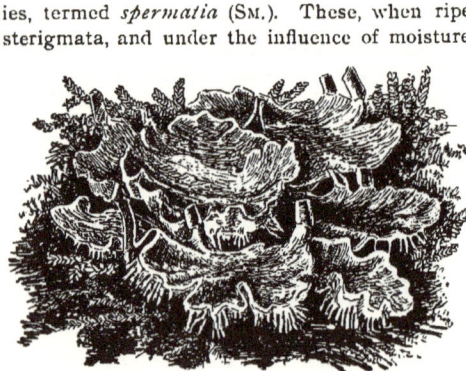

Fig. 10. *Peltigera canina.* Natural size. (After Kerner.)

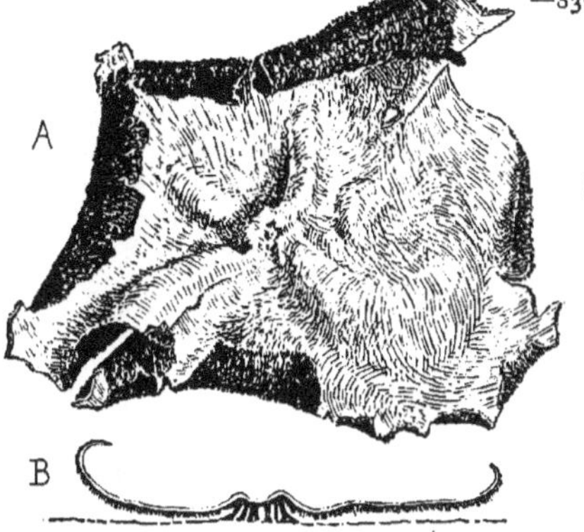

Fig. 11. *Umbilicaria Dillenii.* Natural size A, view from above; B, diagrammatic vertical section, showing method of attachment to substrate. (Original.)

Fig. 12. *Ramalina calicaris.* Natural size. (After Rabenhorst.)

which subsequently gives rise to the spore-producing fruit. In such lichens as *Parmelia*, however, the most careful search has failed to discover any trace of a female organ, and there is other evidence that the spores arise in an entirely non-sexual manner. Moreover, the sper-

Fig. 13. *Usnea barbata.* Natural size. (After Sachs.)

matia of such lichens have been found to germinate and produce a mass of hyphæ entirely similar to that grown from spores. The conclusion seems warranted, therefore, that the spermatia of the majority of lichens have in the course of evolution changed their function, and while they were orginally male reproductive bodies, they now serve as supplementary non-sexual spores.

Having studied in detail one typical example, it remains for us to consider the principal modifications of form which the different parts of lichens exhibit.

The chief forms of thallus are briefly indicated in the following table:

I. Closely united with the substrate, so as to appear like an incrustation; without rhizoids. (Fig. 9)..CRUSTACEOUS
II. Attached to the substrate by rhizoids or by definite portions of the lower surface; lobes numerous or ample...FOLIACEOUS
    1. With numerous rhizoids or points of attachment.
       a. Lobes lying close to the substrate. (Fig. 1)........ADNATE or APPRESSED
       b. Lobes ample and ascendant. (Fig. 10)..................FRONDOSE
    2. With a single point of attachment near the center. (Fig. 11)......UMBILICATE
III. Arising from a single point of attachment, and growing more or less perpendicular to the substrate; branched and shrubby or pendulous, the branches flattened. (Fig. 12) or terete (Fig. 13) ....FRUTICULOSE
IV. Possessing both a horizontal and a vertical part, the former being crustaceous or foliaceous, and the latter consisting of individual members, called *podetia,* that may be goblet-shaped (Fig. 14, A), club-shaped, or cylindrical (Fig. 14, B), and either simple or branched...CLADONIÆFORM

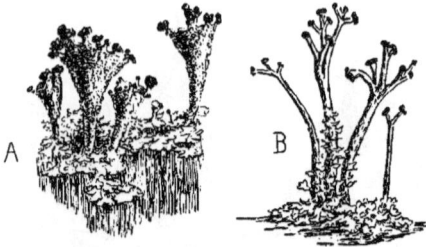

Fig. 14. A, *Cladonia pyxidata;* B, *Cladonia cristatella.* Natural size. (Original.)

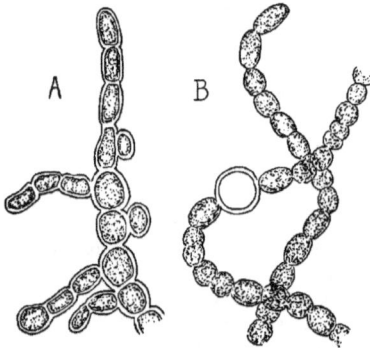

Fig. 15. A, gonidia of *Graphis scripta*; B, goni- mia of *Leptogium.* Magnified about 250 diameters. (Original.)

As regards the charac- ter of its surface, the thallus may be smooth, with a bloom (*pruinose*), powdery (*pulveru- lent*), mealy (*tartareous*), scufy, warty (*verrucose*), hairy (*tomentose*), cracked (*rimose*), covered with a network (*reti- culate*), divided into small, regular spaces (*areolate*), or with indentations or depres- sions (*lacunose*).

In texture the thallus may be thin and p a p e r y (*membrana- ceous*), moderately firm (*cartila- ginous*), o r t o u g h like leather (*coriaceous*).

Sometimes the soredia, in- stead of being mere granular heaps, become coral-like projec- tions, and are then called *isidia*.

The different kinds of *Algae* which serve as g o n i d i a are mostly e i t h e r grass-green or bluish green. When bluish they are termed *gonimia.* Some of the commoner forms of gonidia are shown in Figs. 3 and 15. In a few cases gonidia occur in the hymenium, and are disseminated with the spores.

The principal forms of apothecia are as follows:

I. Hymenium exposed when mature....... ....GYMNOCARPOUS
   A. Hymenium solid at maturity.
      1. Rounded in outline, concave, flat, or convex.
         a. Disk margined, at least when young, by a *thalline exciple*—that is, one
            which is continuous with the thallus and the same in color. (Figs. 2
            and 4) ....SCUTELLIFORM
         b. Disk margined only by a *proper exciple*—that is, one which is a con-
            tinuation of the hypothecium. and which does not contain gonidia.
            *Saucer-shaped or shield-shaped, with the exciple distinct at the
              margin. (Fig. 16, A) ....PATELLIFORM
              †Exciple coal-black............................. ....LECIDEINE
              ‡Exciple paler than the disk.................... ....BIATORINE
              §Strongly convex or globular, the exciple at length covered or
              obscured by the disk. (Fig. 16, B)................ CEPHALOID
         c. Disk margined by both proper and thalline exciples. (Fig. 16, C)
                                      ZEORINE
      2. Elongated in outline, furrow-like, straight or curved, simple or branching-
         (Fig. 16, D)............. ....LIRELLIFORM
   B. Hymenium becoming a powdery mass of spores by disintegration of the thekes
      at maturity. (Fig. 16, E).... ....CRATERIFORM
II. Hymenium enveloped in a *perithecium*—that is, a proper exciple which is spherical or flask-shaped, and at maturity opens by a spore at the summit, through which the spores escape like spermatia from a spermagone. Within the perithecium is usually another layer (the *amphithecium*) which gives rise directly to the hymenium. (Fig. 16, F)
                                            ANGIOCARPOUS

Spores may be colorless* or colored (mostly brownish or olive). The
typical forms of spores are illustrated in Fig. 17. As may be seen also in
these illustrations, spores may be *simple,*—*i. e.*, consisting of but a single cell,
—(A and C), or they may be divided by partition walls (*septa*) into two, four,
or more compartments (*loculi*), when they are termed, respectively, *bi-*,
quadri-, or *plurilocular* (D, E, I, H, and J). When there are longitudinal
as well as transverse septa, the spore becomes *muriform* (F, G). When
there is a small loculus at each pole (as in B) the spore is termed *polar-bilocular.*

A few lichens have spores almost large enough to be seen with the naked eye: the majority are microscopic and in many cases exceedingly minute.

The spermagones of most lichens so nearly resembles the type as shown in *Parmelia* that they will be easily understood without further description.

The size of spores is commonly expressed in terms of the microscopical unit known as a *micromillimeter*, which is equivalent to one thousandth of a millimeter, and is indicated by the abbreviations *mic.*, *mm.*, or by the Greek letter micron, μ. It is customary to write the length as the numerator of a fraction and the breadth as the denominator, and to indicate the minimum and maximum of each dimension. Thus for the spores of *Parmelia conspersa,*

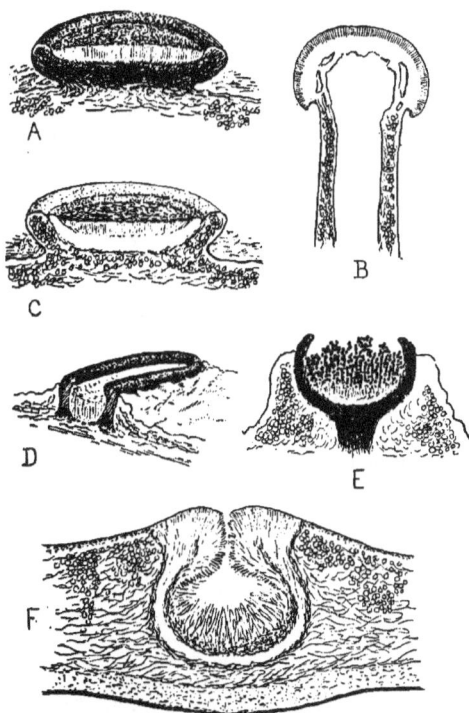

Fig. 16. Forms of apothecia. A, patelliform (*Lecidia*); B, cephaloid (*Cladonia*); C. zeorine (*Lecanora*); D, lirelliform (*Graphis*); E, crateriform (*Acolium*); F, angiocarpous (*Endocarpon*). Variously magnified. (Original.)

---

*According to Professor Tuckerman, elongated spores are typically colorless, while the broader forms are typically colored. Broad spores which are without color he calls *decolorate.*

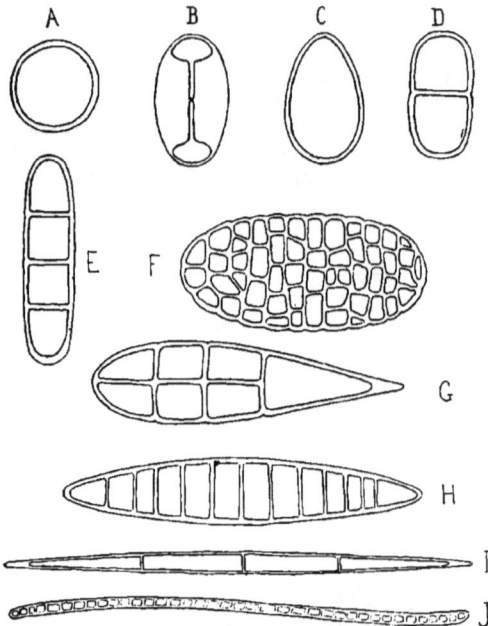

Fig. 17. Forms of spores. A, globose; B, ellipsoid; C, ovoid; D, oblong; E, dactyloid; F, oblong-ellipsoid; G, cymbiform; H, fusiform; I, acicular; J, cylindrical. A and C are simple; D, bilocular; E and I, quadrilocular; H. and J, plurilocular; B, polar-bilocular, F and G, muriform. (Original.)

which vary from .007 to .012 millimeters in length and from .005 to .007 in breadth we should write the expressions $\frac{7\cdot12}{5\cdot7}$ mic. or μ.

As soon as possible after an unknown specimen is collected it is desirable to examine the spores. Having made a thin section of the apothecium or removed a small portion of the hymenium, it may be treated with a little potassic hydrate on the slide and crushed somewhat under the cover glass. Iodine is sometimes useful as a coloring agent. It gives a yellowish or brownish tinge to protoplasm and turns the thekes b l u e. The microscope should be provided with an eyepiece micrometer and the value of the divisions should be carefully ascertained for each objective. Thus equipped it is an easy matter to measure the spores, and these measurements should be recorded on the label of the specimen together with a drawing of a spore, showing outline, number of septæ and (by shading) the presence of color. Cambridge, Mass.

(To be continued.)

WHAT TO NOTE IN THE MACROSCOPIC STUDY OF LICHENS II.

BRUCE FINK.

VARIATION IN LICHENS.

Before passing to a consideration of the gross morphology of the fruits of lichens, it may be stated that lichens are scarcely more varied as to form, size and color than many undoubted morphological units of the plant kingdom, and that the variations are by no means so great that these characters can not be depended on in the description and determination of species and varieties. The student will find in some works on lichens that, in an attempt at brevity, there has been an omission of some characters that are very essential to successful determinations and that in others the same end is gained by a dogmatic statement regarding form, size and color, which would lead the beginner to expect in the plants a like constancy, which by no means exists.

THE APOTHECIUM.

Likewise in the fruit, or apothecium, the main features of gross morphology are size, form and color; and these will now be taken up and considered, one at a time. The apothecia are usually superficial and large enough to be seen easily with the unaided eye. But in some instances they are so small that they can be made out with difficulty with the hand lens. Or they may be immersed in the thallus and indicated at the surface by a slight elevation or depression as a disk or ostiole, or they may, when immersed, be scarcely discernable in any way except in sections of the thallus. From .1 to 5 mm. is well within the range for diameters of apothecia.

The apothecia are most commonly saucer-shaped or some slight modification of this form as when the disk is flat or somewhat convex instead of concave. In some instances the disk becomes very concave, when finally the apothecium may be called cup-shaped, and in others it is strongly convex, finally giving the apothecium a more or less spheroidal form. In all of these forms, the outline of a transverse section of apothecium would usually be very nearly a perfect circle; but the form may become irregular as growth proceeds, so that at maturity this outline is quite irregular. In other lichens the apothecia are of some other form from the beginning. Thus there are the elongated and often branched forms such as are found in *Graphis*, and the difform or variously irregular forms as in *Arthonia*. Again, some apothecia are produced into a well developed perithecium, and these usually approximate a spherical form.

THE DISK.

In those lichens in which the exciple is not produced into a perithecium, the upper surface of the apothecium is naked, except for a very thin film of thallus, which may persist as an epithecium, a structure not mentioned in the descriptions of species. This upper and essentially naked surface, whether flat or more or less strongly concave or convex, forms the disk. The outline of the disk, then, may be circular or variously elongated or irregular, varying in this respect with the form of the apothecium as a whole. In color the disk varies considerably even in the same species. It is usually light colored in its early development and commonly becomes darker as it reaches maturity. The final color, then, may be a light or darker flesh-color or a light or darker shade of yellow, orange, red, brown, chestnut, olive, or even black. And the color is very seldom the same as that of the thallus, but the surface may be pruinose with a usually white powder, concealing the essential color of the disk.

THE EXCIPLE.

Below the disk is the hymenium, which may easily be made out in sections with the hand lens. This structure is usually lighter in color than the disk and is composed of paraphyses and asci. Below the hymenium is the hypothecium, often darker in color than the hymenium above it, so that the line of demarkation between the two structures may easily be made out with the hand lens. The hymenium and the hypothecium are hardly to be studied macroscopically, and are mentioned here mainly that another structure, the exciple, may be located with reference to these two structures. Then the exciple may be said to form the saucer-shaped or cup-shaped covering around the hymenium, being an upward continuation of the hypothecium on all sides. Such rather is the *proper exciple*, but there is sometimes outside of this, or more often replacing it, what is known as a *thalloid exciple*. This last is similar to the thallus in structure and usually of the same color, while the proper exciple is never of the same color as the thallus and usually approximates in color the disk.

Either of the exciples may be permanently absent, and either or both may be quite evanescent and only to be seen in young apothecia. But usually one is permanent or tardily disappearing so that it may be seen readily with the eye or the lens, and the nature and degree of development and permanence is of considerable value in the classification of lichens, even to the determination of species. The perithecium has been mentioned, and this is simply a produced exciple found in some lichens, and growing completely around the upper part of the hymenium, except the small opening or ostiole at the summit. The margin of a proper exciple is usually about at the level of the outer margin of the disk, or it may be somewhat raised above the disk. This margin is almost always quite entire, while the margin of a thalloid exciple is frequently crenulate, crenate, variously branched, ciliate or irregular.

POSITION OF THE APOTHECIA.

Perhaps the only thing yet remaining to be said regarding the fruit, is something concerning its position with reference to the thallus. Sometimes the apothecium is raised on a slender upward extension of the thallus, a short stalk or pedicel, quite different from the podetium of a *Cladonia* or the stipe of a *Calicium* and most frequently met in the larger foliose lichens. The stalk may be absent and the apothecium attached to the thallus at the centre of the ventral side of the apothecium. Such apothecia are said to be sessile. Again, the apothecium may be more closely attached to the thallus by all of its lower side, and then it is said to be adnate. Finally, the apothecia may be more or less immersed in the thallus, sometimes deeply, so that, when the disk is more or less over-grown by the thallus or by a perithecium, the structure is often quite obscured.

The developement of the apothecium begins below the surface of the thallus, and the tendency in general is to become more and more superficial as maturity is reached. So it comes about that the apothecium sometimes remains permanently more or less immersed, or more commonly becomes superficial or stalked. And somewhat varying conditions as to position with reference to the thallus may be expected in many species.

STIPES AND PODETIA.

Very naturally, we may consider next in order podetia and stipes. Both are to be regarded as structures developed originally for the purpose of supporting the spore-bearing organ, or apothecium, and raising it up into the air. But in the *Cladonias* and *Stereocaulons*, a secondary function has replaced the original to some extent, and the original stipe, a structure devoid of algal cells as that of the *Caliciums*, *Cypheliums* and *Coniocybes*, has in these first two genera taken to fostering the algal cells, thus becoming a true thallus, whose vertical expansion and often extensive branching greatly increases the area of surface below which the algae may find protection. Thus the stipe differs from the podetium in that the former is devoid of algal cells and the latter not, the former belonging to the fruiting tract and the latter, doubtless by modification, more properly to the vegetative tract.

RHIZOIDS AND CILIA.

And after thus considering the gross morphology of the thallus and the fruiting tracts, there are a number of less conspicuous structures, less constantly present even than the fruit, or not so often to be seen with eye or lens, but still sometimes quite apparent. First among these may be mentioned the rhizoids and cilia. The rhizoids are found on the ventral side of most foliose thalli and serve as attaching organs. They appear to the eye as root-like bodies, varying in color from white to black. We say root-like, but do not call them roots, since they differ from true roots very much as to minute morphology. The cilia are like the rhizoids in structure, but are found on the upper surface of the thallus or along the margins.

The hyphal rhizoids of the crustose lichens are quite different morphologically, but these need receive no attention here on account of their minute size. However, it is in order to state that the functions of cilia are to retain drops of water and gradually absorb them, and sometimes when quite numerous, to protect against cold and dust. Closely related to rhizoids is the simple attaching organ on the ventral sides of the thalli of *Gyrophoras*, *Umbilicarias*, and many *Endocarpons*. This structure is known as the umbilicus.

SOME OTHER STRUCTURES.

Spermagonia, soredia, cephalodia and cyphellae are structures which occur on the surface of thalli. The spermagonia are organs of doubtful nature and function, sometimes quite conspicuous as dark colored spots on the upper surface of the thallus as in some *Parmelias* and other large foliose lichens; but these structures are more often minute and of the color of the thallus so that they appear only in sections. They were formerly thought to be of considerable value in the determination of species. The soredia are little powdery masses, usually whitish in color, and scattered over the surface of the thallus as in *Pyxine sorediata* and many other lichens. Cephalodia are wart-like bodies found on the upper surface of some lichens as in *Peltidea aphthosa*, or within the thallus, as in some other lichens. Cyphellae are small pits or depressions in the lower surface of some foliose lichen thalli as in some *Stictas* and *Stictinas*. The further consideration of the various structures and their functions belong rather to minute morphology and to physiology.

Finally in some lichens the so-called hypothallus is conspicuous to the eye or with the lens. This is true of some members of the genus *Pannaria*. For instance, in *Pannaria nigra*, this structure appears as a bluish-black ring all around the thallus. Its nature is not well understood, though possibly it is a remnant of some lichen that the *Pannaria* has over-grown. Some of the older authors considered the rhizoids a portion of the hypothallus.

CONCLUSION.

In these days of microscopes and microscopic study, there is some danger that the beginner will actually come to think that there is little to be seen in lichens and other plants below the spermaphytes without using a microscope. The object of the present paper is, then, to call attention to the very many features of lichen-morphology which may readily be observed in the field, with no other aid than the eye, or this supplemented by a good hand lens. The points considered in the above pages are well known to lichenists, but we do not know where they have previously been brought together in compact form so that the beginner may have them for ready reference.

It has not been possible to give, in so brief space, every feature of the gross morphology of lichens and all variations, but the statements are intended rather to be suggestive. It is believed that, if the student will repeatedly read the suggestions given, and then observe more carefully than ever before, he will soon become convinced that the lichens have more definite features of gross morphology than he had previously supposed, and that these features require careful attention at his hands. It is well known that a number of the readers of ,the BRYOLOGIST are becoming interested in lichens, and it has seemed appropriate that such statements as are given herein should be brought to the attention of these persons as an aid to careful observation. ⸻ ⸻ Grinnell, Iowa.

BOOK REVIEWS.

SECOND EDITION DIXON & JAMESON.

J. P. NAYLOR.

A somewhat careful comparison of the last edition of Dixon & Jameson's Student's Handbook of British Mosses, with the Manual, Barnes' Keys, and the file of the BRYOLOGIST, shows that, of the six hundred and two species described in the Handbook, four hundred and thirty-three are American forms. This is a little more than seventy-two per cent. Of the one hundred and fifteen genera all but six are found in the United States, and of these six, only one has more than one species. The British Moss flora seems to be particularly rich in one species genera; there being forty-four of them which contain but a single species.

The fact that the genera are so largely American, taken in connection with the plates and the admirable descriptions, makes the book most valu-

able to the beginner in moss study. It is often more difficult for the novice
to settle upon the genus than it is to determine the species after the genus is
found. Whether a specimen is an Anomodon or Leskea, or whether a Barbula or Tortula, are good examples of this sort of trouble. The expert does
not see why, of course, but the beginner does, as some of us could testify
from past experience. It is true that there are scores of American species
and many genera not found in the work, but the most commonly occurring
ones are described, and the smaller number of prominent forms is a help
rather than a disadvantage. The only objection to the book is its price,
which is something over six dollars to American students.

Greencastle, Indiana.

GROUT, A. J. Mosses with a Hand-Lens, a Non-Technical Handbook of
the More Common and More Easily Recognized Mosses of the North-
eastern United States. Second Edition, Revised, Enlarged and Including the Hepatics. 8vo. pp. xvi + 208, 151 figures in text. Published by
the author and by the O. T. Louis Company, New York City.

The first edition of this book was reviewed by Dr. Best (BRYOLOGIST,
4:28. 1901). who described its purpose and general character. The call for
a second edition within so short a time shows clearly that there is a demand
for a work of this nature, and that the first edition has been of distinct service to students beginning the study of bryology. The second edition follows the same general plan as the first, but there are a number of noteworthy
additions, In the first place, about fifty species of mosses, not included in
the first edition, are described and figured. In the second place, certain
characters are made use of which can be discerned with the hand-lens by
making preparations on glass slides, similar to those required for the compound microscope. In this way the features of the leaf-margin, the length
of the costa, certain peristome characters, the position of the reproductive
organs, and even the general peculiarities of the leaf-cells are consistently
included in some of the descriptions.

The most important addition, however, is found in the portion of the
work devoted to the Hepaticae, which are treated in the same way as the
mosses. As there is no other popular work dealing with our American liverworts, the present chapters will be especially welcome. Fifty-four species,
representing forty genera. are described and the majority of these species
are figured. With the exception of Fossombronia, Marsupella and Diplophylleia, all of our more common genera are included. The author empha-
sizes the fact that many of our hepatics can be identified, even when sterile,
by characters derived from the purely vegetative organs. To a great extent
this is true, but a satisfactory knowledge of a genus or species can of course
be secured only by the study of fruiting specimens. It is to be hoped that
Dr. Grout's book will arouse interest in the hepatics, as it has already done
in the mosses, and that collectors will devote more attention to these neg-
lected plants. Even in the East our knowledge of the distribution of common species is still far from complete. ALEXANDER W. EVANS,

New Haven, Conn.

SULLIVANT MOSS CHAPTER NOTES.

The following names have been added to the list of Chapter Members since July 1st, making the total number one hundred and fifty-three: Mr. Henry Duntun, 139 Franklin street, Jersey City, New Jersey; Mr. W. W. Calkins, 147 California avenue, Chicago, Ill.: Dr. Lewis Sherman, 448 Jackson street, Milwaukee, Wis.: Dr. Bouley de Lesdain, 16 Rue Emmery, Dunkerque, Nord, France.

NOTES.

It is a pleasure to record the fact of the recent appointment of James Franklin Collins as assistant professor of botany, at Brown University, where he has been curator of the University Herbarium since 1894 and instructor in botany since 1899. He is a member of the New England Botanical Club, Josselyn Bot. Soc., Sigma Xi., and other societies. He has contributed since 1893 articles in botanical research to Bull. Torrey Bot. Club, Rhodora and THE BRYOLOGIST; also several articles and all the illustrations in W. W. Bailey's "Botanizing" published in 1899.   A. M. S.

NOTE TO MEMBERS OF THE HEPATIC DEPARTMENT.

Though accessions are constantly coming in for our Sullivant Moss Chapter Herbarium of Hepaticae, I would like to arouse greater interest it it. We desire that it become rich in the common species from every state and contiguous district, as new and fuller data on distribution, etc. could then be obtained and, through published lists from time to time, its full contents be made known. Not wishing to merely beg specimens for the Herbarium members can exchange hepatics with me, and I shall be glad to send my list of duplicates to any not already supplied. Correspondence relative to this project and the rarer hepatics will also be gratefully received by me.

CAROLINE COVENTRY HAYNES,
16 East 36th St., New York City.

OFFERINGS.

(To Chapter Members only. For postage.)

Dr. John W. Bailey, "Walker Building," Seattle. Wash. *Porotrichum Bigelovii* (Sulliv.) E. G. Britton; *Claopodium Whippleanum* Sulliv.; *Hypnum uncinatum symmetricum* R. & C.; *H. aduncum tenue* B. & S.

Mr. N. L. T. Nelson, 3968 Laclede ave., St. Louis, Mo. *Brachythecium acuminatum* (Hedw.) Kindb., c.fr. Collected in Missouri.

Mrs. Sarah B. Hadley, South Canterbury, Conn. *Webera nutans* Hedw., c.fr. Collected in South Canterbury.

Mr. J. Warren Huntington, Amesbury, Mass. *Webera nutans caespitosa* Sch., st.; *Grimmia Olneyi* Sulliv., st. Collected in Amesbury.

Mr. Severin Rapp, Sanford, Orange Co., Florida. *Thuidium microphyllum* (Schwaegr.) Best; *Leskea microcarpa* Sch. Collected in Sanford.

Miss Alice L. Crockett, Camden, Maine. *Usnea trichodea* Ach.; *Pannaria lanuginosa* (Ach.) Koerb. Collected in Camden.

Mr. G. K. Merrrll, 564 Main street, Rockland, Maine. *Cladonia reticulata* (Russell) Wainio (C. Boryi Tuck). Collected in Knox Co., Maine.

Mrs. M. L. Stevens, 39 Columbia street, Brookline, Mass. *Cladonia squamosa* (Scop.) Hoffm. forma. Collected in Laconia, N. H.

SPORE DISTRIBUTION IN LIVERWORTS.

W. C. COKER.

(Taken by permission from Botanical Gazette, 37:1, 1904, p. 63.)

It has no doubt been noticed by all observers of the liverworts that, while terrestrial species have as a rule (Riccia and Sphaerocarpus are exceptions) their capsules raised on elongated stalks, furnished either by sporophyte or gametophyte, those which grow on trees seldom elongate their stalks more than enough to free the capsule from the perianth. This difference is plainly due to the fact that the arboricolous species are sufficiently elevated to allow their spores to be well scattered without any special contrivance. It is interesting to note, however, the behavior of the fertile branches of *Porella platyphylla* Lindb. While the vegetative branches of the liverworts remain closely appressed to the bark of the tree the fertile shoots bend away some time before the spores are ripe, and often project a centimeter or more from the substratum. This exposes the spores to the free play of the wind and no doubt prevents many of them from being caught by the leaves of the mother plant. This habit seems to show that even in arboreal forms it may be an advantage to have the capsule removed some distance from the substratum. It will be noticed here that Porella resembles the Marchantiaceae in giving over to the gametophyte the duty of lifting the capsules.

University of North Carolina, Chapel Hill, N. C.

THE ADVANTAGE OF FREQUENT VISITS TO MOSS LOCALITIES.

B. D. GILBERT.

One of the most interesting spots for moss collecting which it has been my fortune to strike is a small swamp lying along the creek and near the head of one of the ponds that line the Sanquoit Valley, Oneida County, N. Y. Lying not more than a quarter of a mile from my home, and easy of access, it is one my favorite hunting grounds; and I seldom go there without finding some species which had not previously attracted my attention.

One of the earliest things found last season was *Camptothecium nitens*. I could scarcely believe in my good luck when I discovered this species, which was recognized almost immediately from one of Dr. Howe's specimens which he sent me many years ago. His specimen from Washington county, and one sent by Judge Clinton from West Bergen swamp, are the only specimens from New York State in the State Herbarium at Albany. I watched this bed of moss all the summer and fall for fruit but none appeared, and so far as can be learned it is seldom found in fruit anywhere. It grows in a solid bed, three to four inches deep, and covering three or four feet of surface, mixed with Equiseta and grasses.

Two mosses are very common in this swamp, and these also grow in solid masses: *Gymnocybe palustris* and *Philonotis fontana*. They are nearly always sterile though both have been found fruiting. Growing with these is another much rarer species—*Hypnum filicinum*, this is most apt to be sterile. In another part of the bog, at the edge of solid ground, grows *Fissidens adiantoides*, mixed with *Selaginella apus* and *Equisetum scirpoides*. Of course in such a favorable spot *Hypnum Schreberi* is met with here and there, and a small form of *Climacium Americanum*, always sterile. *Thuidium delicatulum* is also very abundant. *Hypnum cuspidatum*, which the books report as being a rather rare species, is met with here but like most of the others is sterile. *Amblystegium chrysophyllum* is another barren species of this swamp. *Bryum caespiticium* fruits freely in the spring, but it seems to be about the only species that does so. Perhaps some of our moss experts can explain why the mosses growing here should be so unfertile. The swamp is an open one, filled in summer with *Nephrodium*, *Thelypteris* and *Onoclea sensibilis*, an occasional clump of willows and quantities of *Equisetum hyemale*. Late in the season the ferns have died away and it is then easy to find the plants that grow close to the ground.

A few steps lower down the stream there stands a piece of woods, but the soil underneath is fairly firm, with here and there a damp depression. Many trees of goodly size grow here, while scattered over the ground lie a number of old windfalls. On the bark of the trees are found *Leskea polycarpa*, *L. gracilescens*, *Anomodon minor*, *A. rostratus*, *Amblystegium varium*, *A. adnatum* and *Thelia hirtella*. The other species of Thelia seem to be scarce having been found but once in a piece of hill woods covering a large stone. The logs are covered with *Entodon cylindrothecium*, *E. seductrix*, *Brachythecium lætum*, *B. salebrosum*, *Plagiothecium denticulatum*, and other common Hypnums.

Judging from this locality, it is not safe to conclude that because one has visited and even ransacked a spot once, there is nothing more to be found. Mosses are small and have a way of eluding observation. So it pays to visit the same locality several times in case one or two good things are found at first, as others are liable to be discovered later on.   Clayville, N. Y.

NEW OR UNRECORDED MOSSES OF NORTH AMERICA.

By J. Cardot and I. Thériot.

Translated and condensed from The Botanical Gazette, May, 1904.

Descriptions of new species given in full. See Bryologist, January, March and July, 1905.

Bryum anceps Card. & Thér.

Densely caespitose, yellowish green. Stems simple or sparingly branched, 5-10 mm. high, interruptedly foliate. Leaves in a comal tuft at apex of branches, about 1.5 mm. long, 0.8 mm. broad, concave, slightly decurrent at the broadly ovate base, short acuminate, the median and lower leaves obtuse, the upper subobtuse or subacute, apex entire or subdenticulate; margins revolute at least at one side in the upper half; costa strong, dilated at base, 100μ thick, gradually attenuate and vanishing below the apex; areolation lax, lower cells quadrate or rectangular, the rest hexagonal, about 50μ long, 20μ broad, all the walls thin and soft. Other characters unknown. Plate XIX.

Northwest Montana: In the vicinity of Lake MacDonald, Flathead Co., 1000-2100 m. alt. (J. M. Holzinger and J. B. Blake, 1898).

The relationship of this moss is rather doubtful; it seems, however, to have some affinity with *B. capitellatum* C. Mull. & Kindb., which it resembles in habit and form of the leaves, but it has more slender stems, the leaves somewhat decurrent, revolute in the lower part at least on one side, a looser areolation formed of much wider and softer cells with thinner walls, and a much broader nerve (100μ wide at base, instead of 60μ) ending at a little distance below the apex.

Bryum brevicuspis Card. & Thér.

Synoicous, densely caespitose, lurid-green, Stems 1-2 cm. high, radiculose, branching below the perichaetium. Leaves erect-spreading when moist, spirally twisted when dry, about 1.5 mm. long, 0.65 mm. broad, ovate-oblong, short acuminate; margins revolute from base to apex, denticulate above; costa reddish below, 72-82μ thick, percurrent or often short excurrent; median cells hexagonal or rhomboidal, 30-35μ long, 12μ broad, very chlorophyllose, walls thin, the upper longer, the lower cells larger, laxer, short rectangular, hyaline, marginal cells narrow forming a quite distinct border of two or three rows. Capsule pendulous, short, 1.6 mm. long, 0.9 mm. thick, slightly constricted under the mouth when old and dry, operculum convex-apiculate. Seta flexuous, about 1.5 cm. long. Peristome perfect; cilia apendiculate. Spores smooth, 13-15μ in diameter. Plate XXII.

Missouri: Eagle Rock, on trees (B. F. Bush), 1898).

Easily distinguished from *B. capillare* L. by its synoicous flowers, very briefly excurrent costa and shorter capsule. Differs from *B. provinciale* Philib. by its smaller leaves, more narrowly revolute, with a shorter point not so much denticulate in the upper part, and also by its shorter capsule. It seems more closely connected with *B. Oreganum* Sulliv., but from Sullivant's description and drawings (Exped. Wilkes, Musci, 10, pl. 7, B) the latter has shorter stems, more serrate leaves, not twisted when dry and a longer capsule.

BRYUM SUBDREPANOCARPUM Card. & Thér.

Dioicous, loosely caespitose, green. Stems short, 5-7 mm. high, radiculose below, forming |slender loosely foliate innovations under the perichaetium. Lower leaves remote, upper crowded into a small rosette, 1-1.5 mm. long, 0.6-0.75 mm. broad, ovate-oblong, short-acuminate; margins longly revolute, plane and denticulate in the upper part; costa 80µ thick at base, vanishing usually below the apex; areolation lax, basal cells rectangular, median and upper oblong-hexagonal, 60µ long, 18-20µ broad, marginal cells 1-2 rows narrow, linear, yellowish. Seta reddish below, pale above, 1.5-2 cm. long. Capsule pendulous or nodding, oblong, arculate, with a long neck; operculum convex-apiculate. Annulus broad. Peristome perfect, cilia appendiculate. Spores 8-12µ thick. Plate XXII.

California: Soldier's Home, Los Angeles Co. (Dr. Hasse, 1902; herb. C. F. Baker).

This moss is very near *B. drepanocarpum* Philib., from which it differs by its shorter and broader leaves, denticulate in the upper part, and forming a small rosette or bud at the top of each stem, and by its costa generally vanishing below the apex.

BRYUM CAMPTOCARPUM Card. & Thér.

Monoicous, rather laxly caespitose, yellowish green. Stems short, about 5 mm. high, radiculose, simple or branching. Leaves erect, quite crowded, oblong-lanceolate, acuminate, middle and upper leaves 2 mm. long, 0.6-0.7 mm. broad, the lower shorter: margins incrassate, plane or barely reflexed below, subdenticulate towards the apex; costa reddish, 80-90µ thick at base, shorter in the lower leaves, longer excurrent in the upper leaves: median and upper cells hexagonal and rhomboidal, 54-70µ long, 18-24µ broad, basal cells rectangular, 80-90µ long, 20-30µ broad, marginal cells bi- or tri-stratose, forming a beautiful distinct yellowish incrassate border. Male flower on a separate terminal branch. Capsule nodding, oblong, arcuate, 4-4 5 mm. long, 1 mm. broad, tapering into a neck as long as the sporangium; operculum convex, short apiculate. Seta reddish, 1.5-3 cm. long. Annulus double and triple. Peristome perfect, 0.48 mm. high: membrane of the inner peristome extending more than half the height of the teeth: processes narrowly gaping for a short distance along the keel; cilia 1-3, appendiculate. Spores papillose, 24µ in diameter. Plate XXII.

Newfoundland: John's Beach, wet places (Rev. Arthur C. Waghorne, 1895).

Allied to *B. meesoides* Kindb., *B. drepanocarpum* Philib., and *B. subdrepanocarpum* Card. & Thér. by the form of the capsule, but distinct from these species by the monoicous inflorescence and by the leaves plane on the margins, or nearly so, with a very distinct and thickened border. It also much resembles *B. pallens* Sw. var. *arcuatum* Sch., from which, however, it differs by the shorter leaves, with margins plane or scarcely reflexed at base and denticulate or sinuate at apex.

(To be Continued.)

VOLUME VIII.   **NUMBER 6**

NOVEMBER, 1905

THE BRYOLOGIST

AN ILLUSTRATED BIMONTHLY DEVOTED TO

NORTH AMERICAN MOSSES

HEPATICS AND LICHENS

EDITORS:

ABEL JOEL GROUT and ANNIE 'MORRILL SMITH

CONTENTS

Entered at the Post Office at Brooklyn, N. Y., April 2, 1900, as second class of mail matter, under Act of March 3, 1879.

Published by the Editors, 78 Orange St., Brooklyn, N. Y., U. S. A.

PRESS OF MC BRIDE & STERN, 97-99 CLIFF STREET, NEW YORK

THE BRYOLOGIST

BIMONTHLY JOURNAL DEVOTED TO THE STUDY OF NORTH AMERICAN MOSSES
HEPATICS AND LICHENS.

ALSO OFFICIAL ORGAN
OF THE SULLIVANT MOSS CHAPTER OF THE AGASSIZ ASSOCIATION.

Subscription Price, $1.00 a year. 20c. a copy. Four issues 1898, 35c. Four issues 1899, 35c.
Together, eight issues, 50c. Four issues 1900. 50c. Four issues 1901, 50c. Four Vols. $1.50
Six issues 1902, $1.00. Six issues 1903, $1.00. Six issues 1904, $1.00.

*Short articles and notes on mosses solicited from all students of the mosses. Address manu-
script to A. J. Grout, Boys' High School, Brooklyn, N. Y. Address all inquiries and subscriptions to Mrs. Annie Morrill Smith, 78 Orange Street, Brooklyn, N. Y. For advertising space address Mrs. Smith. Check, except N. Y. City, MUST contain 10 cents extra
for Clearing House charges.*

Copyrighted 1905, by Annie Morrill Smith.

THE SULLIVANT MOSS CHAPTER.

President, Mr. E. B. Chamberlain, Washington, D.C. Vice-President, Mrs. C. W. Harris,
125 St. Marks Avenue, Brooklyn, N. Y. Secretary, Miss Mary F. Miller. 1109 M Street,
Washington, D. C. Treasurer, Mrs. Smith, 78 Orange Street, Brooklyn, N. Y.
Dues $1.10 a year, this includes a subscription to THE BRYOLOGIST.
All interested in the study of Mosses, Hepatics, and Lichens by correspondence are
invited to join. Send dues direct to the Treasurer. For further information address the
Secretary.

THE ONLY MOSS BOOKS

describing the mosses of the North-eastern United States that are now in
print are **Grout's Mosses with a Hand-lens,** and **Mosses with a Hand-
lens and Microscope.**

MOSSES WITH A HAND-LENS, Edition 2, is an octavo volume of 224 pages,
printed on fine coated paper and bound in strong cloth. It describes 168
species of mosses, and 51 of hepatics. Practically every species is illustrated
in the 39 full page plates and 150 other figures.

*It is the only book on mosses included in the catalogue of 8,000 volumes
for a Popular Library issued by the American Library Association (1904.)*

Price, $1.75. A limited signed edition of 25 copies, with actual specimens
of 200 species mounted on interleaved blank pages, is in preparation.
Price $15.00.

NOW is the time to subscribe for MOSSES WITH HAND-LENS AND MICRO-
SCOPE. On and after the date of issue of the fifth part the price will become
$1.25 per part.

SEND FOR SAMPLE PAGES

A. J. GROUT ❧ 360 Lenox Road ❧ Brooklyn, N. Y.

THE BRYOLOGIST.

VOL. VIII. NOVEMBER, 1905. No. 6.

FIG.1.

FIG. 2.

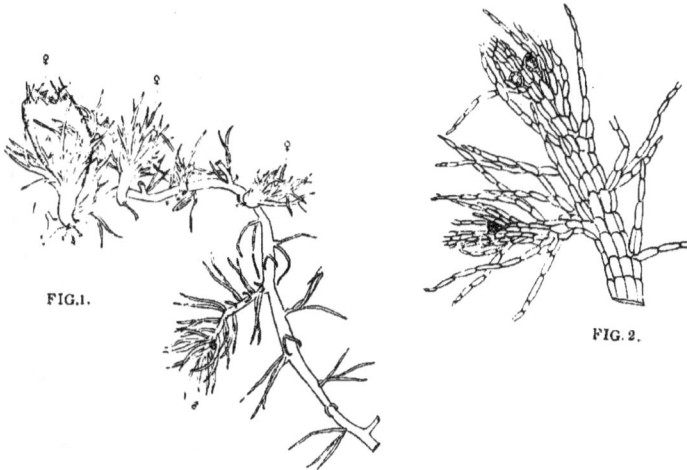

FIG. 1. *Telaranea nematodes longifolia*, M. A. Howe. Postical side, showing archegonia and antheridia. also underleaves X. 22. Reduced ½.
FIG. 2. Antheridial branch showing cells, etc. X. 70. Reduced ½.

TELARANEA NEMATODES LONGIFOLIA M. A. HOWE.

CAROLINE COVENTRY HAYNES.

Telaranea nematodes Gottsche belongs, with the genus, to tropical or warm-temperate countries, South America. South Africa, Cuba and Bermuda. A form of this *T. nematodes longifolia* M. A. Howe (Bull. Torrey Botanical Club, **29**:284, 1902.) is known to occur at a few stations in Georgia and Florida. When, however, Dr. Howe found it growing at Freeport, Long Island, October, 1898, he inclined to the belief that it was not so rare and

The September BRYOLOGIST was issued September 1st, 1905

that it had a wider distribution. The pleasant record is mine of knowing of five stations for it, and thus confirming his belief. I detected it growing among *Pallavicinia Lyellii* (Hook.) S. F. Gray, collected at Arlington, Staten Island, Nov. 28th, 1903, a Field Day of the Torrey Botanical Club, by Mr. W. T. Horne and brought to the New York Botanical Garden. At Highlands, Monmouth Co., New Jersey, I have collected it two consecutive summers. The plants collected in September showed perianths with immature capsules and many antheridia. It was growing among Sphagnum plants; *Cephalozia connivens* (Dicks.) Lindb. in fruit, in the vicinity. These plants were all growing lustily and showed the same shade of tender green and I noticed the Telaranea only from its conferva-like meshes in contrast with the more sharply defined Sphagna. While in North Carolina, February last, I found it at Pinehurst, Southern Pines and at Jackson Springs, all in Moore Co., growing along the borders of running streams, with mosses. These plants showed the ashy-green color of the descriptions. Their delicacy seemed almost ethereal among the larger forms of vegetation and it was wonderful to me that they could survive the winter (for I found old perianths) and start growing as soon as the snow melted away. *Blepharostoma tricho- phyllum* (L.) Dumort, is the only hepatic likely to be confused with it. But *Telaranea nematodes longifolia* has underleaves two or three cells in length, incurved at apices, while the former's approximate the leaves in length. The leaves and underleaves are hair-like, the leaves being five–eight cells long, of a single series of cells to the basal cell. It is autoicous. The archegonia are on short postical branches, the one nearest the apex maturing first. This note is written with the intention of bringing this charming plant to the notice of hepatic students so that they may be on the lookout for it. I shall be grateful if any one finding it, will inform me.

16 East 36th street, New York City.

LICHENOLOGY FOR BEGINNERS—IV.

FREDERICK LEROY SARGENT.

(Begun in May, 1905, issue.)

In the above consideration of the morphology of lichens our attention has been directed particularly to such modifications of form as afford characters useful in systematic classification. But lichens are not only so many species to be named and classified; they are living things adapted to the humble conditions under which they live: and no more interesting field is open to the student of lichens than that which concerns their biology. While a great deal has already been done in the study of the gonidia and their function, and upon the structure and development of other organs, comparatively little has been attempted in the direction of learning the effect of the environment in influencing the form of the different parts or of studying the way these little plants meet the exigencies of their life. Such questions, for example, as the following would, we think, repay careful investigation.

What peculiarity have the dry pasture species in common, and what those of moist woods? Have these peculiarities any such relation to the supply of moisture and light as we find to be the case among higher plants which grow under conditions correspondingly different? What enemies have lichens and how do they protect themselves against their attacks? Are there any special arrangements which facilitate the scattering of spores? How is moisture absorbed by the thallus? Is the moisture which is taken in at one part of the thallus conducted to other parts (or must each part absorb directly through its own surface)? How fast do the different organs of lichens grow? To what extent is the rate of growth affected by differences of moisture, light and temperature? To what age do different sorts of lichens attain? Other questions of a similar nature will readily suggest themselves to anyone interested in these plants, and whereas many problems connected with the minute structure of lichens require considerable technical skill and the use of instruments of great delicacy, all that is needed for the profitable study of questions of the sort above indicated, is intelligent observation and note-taking in the field, or the performance of simple experiments.

A knowledge of the systematic relationship and the names of the species of one's own locality at least is obviously a desirable preliminary to such work. The only manual of North American species is Prof. Tuckerman's Synopsis. (A Synopsis of the North American Lichens: By Edward Tuckerman, Part I, 1882. Published by S. E. Cassino, Boston, Mass. Part II, 1888. Sold by Edwin Nelson, Amherst, Mass. Both parts now out of print). As this work was written for advanced lichenologists, beginners find it difficult to use. The sources of these difficulties are mainly these: first, insufficient acquaintance with the characters of the specimen studied, due to ignorance of just what to look for; second, not understanding the exact meaning of the phrases encountered in the book, or making the necessary allowances; third, the variability and close resemblance of the species themselves. Let us consider how these difficulties may be overcome.

The chief source of trouble is much increased by the too common habit among students of trying to read the descriptions and observe the characters of the species at the same time. Before referring to the book at all, one should make out as far as possible the characters of the specimen in hand. The following schedule of questions will, it is hoped, prove helpful by indicating the important features to be observed in such preliminary examination.

Schedule for Analysis.

What is the locality and habitat?

Is the thallus crustaceous, foliaceous (and appressed, frondose or umbilicate), fruticulose or cladoniæform, or of a form intermediate?

If *crustaceous*, is there a hypothalline fringe, and if so what color is it? Is the surface smooth, pulverulent, tartareous, verrucose, rimose, areolate, or otherwise peculiar? Of what color is it in the younger and in the older portions?

If *foliaceous* or *fruticulose*, is the thallus gelatinous, membranaceous, cartilaginous or coriaceous? Of what form is the margin of the thallus or its lobes or branches? Is the thallus alike on all sides or is there an upper and an under surface? Is the upper or general surface corticate or ecorticate (i. e. with or without a cortex), smooth, polished, wrinkled, channelled, reticulate, lacunose, pulverulent, granular, tomentose, sorediiferous, isidiiferous or otherwise peculiar? What is its color? If sorediiferous or isidiiferous is the whole surface covered or only certain portions of it, and what form do the soredia or isidia assume? If there is a lower surface differing from the upper is it corticate or ecorticate, smooth, wrinkled, pitted, veined, fibrillose (i. e. with rhizoids) or otherwise peculiar? If fibrillose, what is the color of the rhizoids, and are they simple or branched, few or numerous, long or short? Do they extend beyond the margin of the thallus as cilia?

If *cladoniæform*, is the horizontal thallus crustaceous, squamulose (i. e. composed of scale-like lobes or segments), or foliaceous? If crustaceous of what form and color is it, and what is the character of the margin and surface? (See questions given above for crustaceous thallus.) If squamulose or foliaceous, what is the form of the squamules or lobes, what is the form of the margin, and what the color and character of the surface above and below? (See the questions given above for foliaceous thallus.) Of what form are the podetia? Are they solid or hollow, and if branched what is the form and arrangement of the branches? What is the texture and color and what the character of the surface?

[How are the gonidia arranged in the thallus? Of what form and color, are they?*]

Are the apothecia scuttellæform, lecideine, biatorine, cephaloid, lirellæform, crateriform or angiocarpous? Are they immersed, innate, adnate, sessile or stalked? Upon what part of the thallus are they borne?

If gymnocarpous, is the exciple entire at the margin, crenate, ciliate with fibrils or projections or otherwise peculiar? What is the color of the disk when young and when mature, and of what color is the exciple? [Is the hypothecium pale or blackened?]

If angiocarpous, are the apothecia separate and scattered or crowded together and immersed in a common receptacle or stroma? [Of what form is the perithecium, and is it pale or blackened? Is the amphithecium pale or blackened?]

[Are the paraphyses simple or branched? Are the thekes cylindrical, club-shaped (clavate), pear-shaped (pyriform), ovoid, globose, or otherwise peculiar in form? How many spores are there in a theke? Are the spores colorless (pale) or colored? Are they globose, elliptical, ovoid, oblong, cylindrical, fusiform, dactyloid, cymbiform, acicular or of some intermediate form? Are they simple, bi- quadri- or plurilocular, polar-bilocular or muriform? What are the extremes of length and breadth in micromillimeters?]

---

*The questions enclosed in brackets call for the use of the compound microscope, and although here placed in logical sequence among the others, had better be taken up by the student all together at the end.

[What is the form of the spermagones? Where are they situated on the thallus What is the form and size of the spermatia?]

After the student has examined a certain number of species according to the above schedule he will have learned what to look for so well as to have no further need of such help, and will be able to make his preliminary examinations with thoroughness and rapidity. He will learn also that in dealing with certain genera, some features call for more particular observation and some for less. While it is desirable to know the microscopic structure, particularly as regards the gonidia and the spores, it is not always necessary for the recognition of species and even to a less extent of genera, Hence a good beginning may be made with only a hand magnifier, which was indeed all the earlier lichenologists had to aid them.

As regards the second difficulty referred to above, namely, that of not understanding fully the statements of the book or failing to make the necessary allowances, the student will find that these perplexities will disappear in proportion as the mind comes to associate the different phrases with particular features seen in the specimens examined. As in the systematic study of other difficult groups, so with lichens, it is found to be very helpful at first if one can take specimens of which one knows the name, and compare them point for point with the description as given in the manual, for the family, genus and species to which they belong. To enable beginners to do something of this work, there is appended to the present paper a short analytical key by means of which the names of a few of our commonest and most easily recognized species may be determined with tolerable accuracy.

In regard to the third mentioned source of difficulty (the variability and close resemblance of many species) it must be said that even the most advanced students have this to contend with, and as in the case of other perplexing groups, the last resort is the comparison of doubtful forms with authentically named specimens.

Besides Prof. Tuckerman's Synopsis the following writings in English may be profitably consulted by the student:

An Introduction to the Study of Lichens. Henry Willey. New Bedford, 1887.

A Popular History of British Lichens. W. Lander Lindsey. London, 1856.

Guide to the Recognition of the Principal Orders of Cryptograms and the Commoner and More Easily Distinguished New England Genera. Frederick Le Roy Sargent. Cambridge. 1886.

The article "Lichens" in Encyclopaedia Britainnica, Ninth Edition, and the Section on Lichens (pp. 114-126) in Gœbel's Outlines of Classification (Oxford, 1887), give a good general idea of the structure, etc., of these plants.

A Text-Book of General Lichenology, with Descriptions and Figures of the Genera occurring in the United States. Albert Schneider, Binghamton, N, Y., 1897.

Such technical terms as occur in the following Key and have not been defined in the foregoing pages, are used in the same sense as when employed in the description of phanerogams. The abbreviations are the same as those given in Figs. 2-8.

ARTIFICIAL KEY TO SPECIES.*

*The method of using this form of Key may be illustrated by taking as an example a specimen of our typical lichen, with the characters of which we are already familiar. Out of the four alternatives given under "1," we find that only that marked "c," describes our specimen. This refers us to Section 12. Here we find that "c" is the alternative that fits, and this refers us to "15." Under 15 we must choose "a" and thence we go to 16, and from "16a" to "17a" which gives us the name, *Parmelia conspersa.*

1. a. Th. cladoniæform, podetia hollow.......... 2

    b. Th. fruticulose or when young cladoniæform, and the podetia becoming fruticulose by the disappearance of the horizontal th. at least from the base... 5

    c. Th. foliaceous or foliaceous-squamulose... 12

    d. Th. crustaceous..,............... 21

2. a. Ap. brown... 3

    Ap. scarlet or orange 4

3. a. Podetia 3–6 cm. tall, dilated above into shallow cups, bearing apothecia on the rim, and from the centre giving rise to similar stalked cups; very smooth throughout. On the earth.

    Cladonia gracilis var. *verticillata.*

    b. Podetia 5–25 mm. tall, top-shaped, short stalked, the margin spreading, bearing sessile or stalked ap.: granulose, warty or scurfy. On the earth, etc. .... *Cladonia pyxidata.*

    c. Podetia 1–3 cm. tall, goblet-shaped, rather long-stalked and slender, the margin erect, often with tooth-like projections sometimes bearing ap.: cortex disintegrating into a fine glaucous-white powder. On the earth, etc .... *Cladonia fimbriata.*

4. a. Podetia cylindrical, sometimes branched, mostly about 2–4 mm. tall; smooth or with the surface wrinkled. On dead wood, etc.

    Cladonia cristatella.

    b. Podetia elongated-top-shaped, about 15–35 mm. tall; smooth, becoming warty. On the earth.... .... *Cladonia Cornucopioides.*

5. a. Branches cylindrical and hollow........ 6

    b. Branches not hollow.......... 7

6. a. Podetia about 3–6 cm. tall, branches about .5–1 mm. thick, several times forked, awl-shaped at the top, with a smooth firm cortex, brown above, becoming gray in the lower (older) portions. On the earth...... .... *Cladonia furcata* var. *subulata.*

    b. Podetia about 4–10 cm. tall, branches about .5–1.5 mm. thick, the divisions mostly wide-spreading, the sterile tips curved and drooping; ecorticate, the surface fibrous, sometimes appearing mealy or warty: ashy white or tinged with greenish straw-color. On the earth, often forming extensive mats. (Called "Reindeer Moss" from its forming the winter food of that animal.)...... ...*Cladonia rangiferina.*

7. a. Th. softish, cottony within..................................... 8

    b. Th. cartilaginous.. 9

8. a. Branches angular-terete or flattened, often sorediiferous, greenish or straw-colored, sometimes paler below. Mostly on dead wood and (with us) sterile.,............................... *Everina brunastri.*

b. Branches flattened, often isidiiferous, ashy-gray above, paler or often black spotted below. On trees and dead wood (mostly sterile).
Everina furfuracea.

9. a. Branches flattened 10
   b. Branches terete 11
10. a. Branches involute, beset at the margin with a row of dark-colored, finger-shaped projections, 0.3–0.5 mm. long (containing the spermagones). Th. mostly brown above, lighter towards the base where there is sometimes a red stain. Ap. (infrequent) scutellæform, dk. chestnut. On the earth. (The "Iceland Moss" of druggists.)
Cetraria Islandica.
   b. Branches scarcely involute, often with slender projections at the margin, but these are sharp-pointed, yellowish or gray like the th., usually exceed 1. mm. in length and do not contain spermagones. Ap. scutellæform; dk. brown orange ; th. ex. often radiately fibrilose On trees and rocks.... ......*Theloschistes chrysophthalmus.*
   c. Branches not involute, without spermagone-bearing projections differing in color from the th. which is pale greenish or straw-color, rather rigid, more or less reticulately-lacunose and quite variable in the form and number of its divisions, Ap. scutellæform; dk. pale, not differing much from the th. in color. Mostly on trees.
Ramalina calicaris.
11. a. Th. greenish, covered with numerous fibrils of the same color: medulla consisting of· a cottony layer surrounding a tough central cord. Ap. scutellæform: dk. pale; th. ex. radiately fibrillose. On trees (called "Bearded Moss")............. ..... *Usnea barbata.*
    b. Th. dark brown, smooth, sometimes with pale soredia: medulla uniform throughout. Ap. (rare) scutellæform, small. Mostly on trees and dead wood... ....*Alectoria jubata.*
    c. Th. ashy-gray, the branches clothed with granules which may become coralloid or scale-like: medulla firm and uniform throughout. Ap. cephaloid, dark brown or black. On rocks or on the earth.
*Stereocaulon.**
12. a. Th. umbilicate.. 13
    b. Th. frondose 14
    c. Th. appressed, the margin sometimes ascendant 15
13. a. Th. cartilaginous, pale brown or ashy above, fawn-color to dark-brown below, smooth on both surfaces, lobes sometimes much crowded and overlapping. Ap. angiocarpous, imbedded in the th. On rocks, near water so as to be occasionally submerged.
Endocarpum mineatum.
    b. Th cartilaginous, ashy-color above, whitish toward the center; below pale brownish or ash-colored: smooth on both surfaces, often pruinose; with numerous pustular protrusions above, having correspond-

*Of genera thus marked we have several species, the discrimination of which is too difficult to be considered here.

ing indentations below. Ap. sub-scutellæform (appearing as if lecidiene), often clustered. On rocks in dry situations.

Umbilicaria pustulata.

c. Th. coriaceous, often very large, brown above, smooth and even; below 'intensely black with crowded short fibrils. Ap. (infrequent) much as in the last but with dk. ridged concentrically, On rocks in rather dry situations.............................*Umbilicaria Dillenii.*

14. a. Th. cartilaginous, lobes rounded, numerous and crowded, lacunose, pale grayish-green above, whitish or here and there blackening below; no fibrils at the margin. Ap. sub-pedicellate attached obliquely to the margins and summits of the lobes; th. ex. thin, entire or crenate. Sp. simple, ellipsoid. On trees and dead wood.

Cetraria lacunosa.

b. Much as in the last but with marginal fibrils and crenulate exciple. On trees and dead wood.... ....*Cetraria ciliaris.*

c. Th. coriaceous, with rounded sinuses, strongly lacunose-reticulate, tawny or olivaceous; paler below, filbrillose in veins around naked spots. Ap. (infrequent) sessile at the margin of the lobes; th. ex. entire. Sp. cymbriform, 2–4-locular. On trees and rocks. (Formerly esteemed as a pulmonary medicine from its resemblance to a lung).......... ....*Sticta pulmonaria.*

d. Th. coriaceous, with rounded (sterile) lobes and narrower erect ones bearing at the end shield-shaped (often revolute) apothecia; thallus mostly veiny below and with long rhizoids. Ap. innate on the upper surface of the lobes; th. ex. torn-crenate when young. Sp. fusiform-acicular, 4-locular. On the earth or on rocks, mostly among mosses.......................... ....*Peltigera.**

e. Th. cartilaginous-membranaceous, with rounded lobes, not veiny below, and with short rhizoids. Ap. as in the last but reniform and borne on the under surface of the lobes. Sp. fusiform-ellipsoid, 4-locular, Mostly on rocks and trees...................*Nephroma.**

15. a. Th. not gelatinous when moist........ 16
b. Th. gelatinous when moist.............. 20

16. a. Th. pale green or straw-colored above, blackening below....... . 17
b. Th. grayish or whitish ash-colored above, blackening below....... 18
c. Th. grayish or whitish ash-colored above, fawn-color or whitish below...................... 19
d. Th. olivaceous brown or bronze-colored above, blackening below and with black rhizoids toward the centre, membranaceous, closely appressed. Dk. chestnut; th. ex. crenulate. Sp. simple, ellipsoid, colorless. On trees and rocks.................*Parmelia olivacea.*
e. Th. pale yellow to bright orange above, white below, rather loosely appressed. Dk. orange; th. ex. entire. Sp. polar-bilocular, ellipsoid, colorless. On trees and rocks near large bodies of water.

Theloschistes parietinus.

17. a. Th. cartilaginous-membranaceous, the lobes mostly rather narrow,

sub-linear and much divided, smooth, not wrinkled. Dk. chestnut; th. ex. entire. On rocks.....................*Parmelia conspersa.*

b. Th. cartilaginous: the lobes mostly broad and rounded, with numerous distinct wrinkles on the older portions. (Not commonly fertile; dk. as in the last, th. ex. crenulate or sorediiferous). On trees and rocks.... .. ..*Parmelia caperata.*

18. a. Th. commonly reaching a diameter of 10-20 cm, or more, rather loosely adherent: lobes flat or concave, repand, rather narrow, becoming reticulately rimose above, densely black fibrillose below, the rhizoids reaching the margin. Dk. chestnut; th. ex. rather thick, sub-crenulate. Sp. simple, ellipsoid, colorless. On rocks.

Parmelia saxatilis.

b. Th. smaller, closely adnate: lobes flat, smooth, rounded and crenate or more deeply divided, rhizoids black and extending to the margin, but not prominent. Dk. and sp. much as in the last: th. ex. mostly thin and entire. On trees and rocks......*Parmelia tiliacea.*

19. a. Th commonly reaching a diameter of 10–20 cm, or more; lobes rounded, flat or concave, somewhat ascendant; becoming reticulate, rugose and often with soredia or isidia. Dk. chestnut; th. ex. crenulate. Sp. simple, ellipsoid, colorless. On rocks and trees.

Parmelia Borreri.

b. Th. se'dom more than 5 cm. broad, lobes sub-linear, convex, often overlapping and appressed, smooth, without soredia. Dk. brownish-black or gray-pruinose when young: th. ex. mostly entire. Sp. bilocular, ellipsoid, brown. On trees, dead wood and rocks.

Physcia stellaris.

20. a. Th. mostly dark olive green, without distinct cortical layer. Ap. scutellæform ....*Collema**

b. Th. mostly lead-colored, with a distinct cortical layer. Ap. scutellæform, zeorine or biatorine........................,*Leptogium.**

21. a Ap. scutellæform or æeorine............. 22

b. Ap. cephaloid, stalked: dk. rose-pink or flesh-colored; proper exciple (which constitutes the stalk) pinkish-white. Sp. simple, fusiform-oblong, colorless. Th. granular, ashy-gray. On the earth.

Bæomyses roseus.

c. Ap. lirellæform, branched, innate, black, with or without a th. ex. Sp. ellipsoid-fusiform, pluri-locular, colorless. Th. inconspicuous, forming a very thin, smooth, whitish incrustation on bark.

Graphis scripta.

d. Ap crateriform, sessile, with a black dk. and proper ex., surrounded by an accessory thalline one. Sp. ellipsoid-oblong, bilocular, dark colored. Th. greenish-yellow, granular. On dead wood.

Acolium tigillare

22. a. Th. drabish-white, becoming rimose or wrinkled. Dk 1-3 mm. broad flesh color or pinkish pruinose; th. ex. thick, entire. Sp. simple, ellipsoid, colorless $\frac{50\cdot90}{23\cdot40}$ mic. On bark.. ....*Lecanora pallescens.*

b. Th. much as in the last but often granular. Dk. about .5–1. mm. broad, redish to dark brown; th. ex. rather thin, entire or crenulate. Sp. as in the last but $\frac{9-20}{7-11}$ mic. On bark, dead wood and rocks.

Lecanora subfusca.

c. Th. areolate verraculose, pale-greenish, yellowish or whitish. Dk. about .2–.8 mm. broad, pale yellowish, buff or ochraceous-brown; th. rather thin, entire or crenulate. Sp. as in the last but $\frac{9-16}{4-7}$ mic. On bark, dead wood and rocks ....*Lecanora varia.*

THE END.

Cambridge, Mass.

SOME COMMON ERRORS.

EDWARD B. CHAMBERLAIN.

Several times during the past year I have noticed that some of the members of the Sullivant Moss Chapter were abusing the assistance given them by professional bryologists. Usually, the fault was in the misuse of the words "determined by," when sending out specimens offered through the BRYOLOGIST.

I have recently received from Chapter Members mosses which were labelled as named by this or that specialist, when the date of the collection and the often undried condition of the material made it impossible for the specialist to have seen either the specimen or even a duplicate of it. Further inquiry showed the facts to be somewhat as follows: A, collected a moss, and sent it to B, who determined it. Then A, wishing to offer it to the Chapter, and not having on hand a sufficient quantity, went to the spot where he previously found the moss, or where he now thinks he found it, secured more material which looked like the same thing, and distributed this last, labelling it "detr. B," *although B had actually seen none of the second collection.* This second collection may have been the same as the first, but it very probably was not. Under any circumstances it was very unjust and discourteous to the person naming the original collection to make him sponsor for the second. It is really a forgery of his determination.

To avoid such errors the following rule should be adhered to in the use of the words "determined by" or "verified by." Never under any circumstances mark a specimen as determined by another person than the collector unless that person has actually seen either the specimen itself or a true duplicate of it. In the latter case it is very much better to use the words " duplicate determined by." In such a case as that outlined above, the specimens must be marked as determined by the collector, for nobody else had anything to with them.

In this connection it may be well to explain what a duplicate is. By duplicates in mosses are meant, strictly speaking, specimens of the same species, collected in the same locality, upon the same substratum, by the same person and on the same date. The strictness with which this is to be interpreted depends to some extent upon the species of moss in question. In the case of such genera as *Ulota* or *Grimmia,* or of certain of the

Hypnaceae, where several species may be mingled in the same tuft, by far greater care is to be used than in the case of say *Mnium hornum*, which is usually found in pure tufts.

Under all circumstances, the members of the Sullivant Moss Chapter should take especial pains in the matter of the specimens offered by them for distribution. First, be sure the specimens sent out are real duplicates, not a lumping of two or three different collections which you "guess" are the same. Secondly, when having specimens determined by another, send large quantities; if you are not familiar with the species send *all* the material you propose to use in the distribution. Thirdly, be scrupulously exact in citing the authority for the determination; don't endanger another's reputation by your own carelessness.   Washington, D. C.

ENCALYPTA PROCERA BRUCH.

E. J. HILL.

The finding of a sterile Encalypta on rocks at Lockport, Ill., in 1904. led to a correspondence with Mrs. E. G. Britton regarding its specific character In 1889 I had collected *E. procera* in fruit on Presque Isle, Marquette, Mich The question arose whether this might not be *E. streptocarpa* Hedw. (*E. contorta* (Wulf.) Lindb. as now called from an older specific name), or some other member of the genus. The specimens from Presque Isle consisted of one fruiting and a few sterile stems found in a tuft of *Distichium capillaceum*. The peristome of the capsule was unfortunately injured while examining it, but enough remained combined with other characters to lead to its identification as stated. A recent comparison with the barren stems of *E. procera* from British Columbia and fruiting plants of *E. contorta* from two stations in Europe sent by Mrs. Britton, which have a different capsule, strengthened this conclusion. A further search led to the detection of antheridia on the stem of the fruiting specimen showing that it is monoecious as is the case with *E. procera*. The Distichium with which it grew was abundantly fruiting, some barren stems of *Myurella Caryana* were also in the tuft.

The specimens from Lockport being sterile, the vegetative character can only be used to determine the species. Comparing the leaves the same type is seen in the plants from the three American localities. The leaf of *E. contorta* is longer and relatively narrower than in *E. procera* which is broader towards the top, more decidedly lingulate, sometimes subspatulate. The costa of the latter usually ceases further from the apex. The cells of the basal hyaline part are larger especially near the costa than those of *E. contorta* and at times approach a square form. The leaf characters of the two are well shown in the figure in Roth's Europäischen Laubmoose Taf. XLI. From the evidence the specimens best agree with *E. procera*.

The stems of these mosses are well provided with propagula. It was noticed at the time of identifying the plants from Presque Isle, and a reference in a note under *E. procera* in Schimper's Syn. Mus. Eur. states that simple and branching filaments resembling prothallia occur in the axils of the leaves. These structures are fully treated by Correns (Vermehrung der Laubmoose, p. 97-101 : 1899.) mostly under *E. contorta* as those of *E. procera* are essentially the same. Comparing three collections made at Lockport, August and November, 1904, and June, 1905, I find them most abundant on the November plants, least on the June, seeming to increase as the season of growth advances, and as Correns states, ripening in the fall. They are mostly found above the middle of the stem, but not quite to the top in the freshly growing part. They are generally branched or forked, single or in tufts, the lower part smooth, brown or chestnut, divided by rather distant oblique partitions, the upper or propagative part comprising the brood-body chlorophyllose, papillose, the partitions vertical to the outer wall. The partitions plainly show that they are transformed rhizoids though short simple ones may be chlorophyllose almost from their origin, the brown portion very slight. They spring from the stem and the basal part of the costa. None have been observed on the lamina. The cells of the papillose part are mostly longer than wide commonly 25μ long by 22-30μ in diameter, but varying between 20-80 x 20-55μ. Rhizoids which function as such may sparingly branch from the propagula but are much smaller, 5-6μ in diameter.

In *E. contorta* as described by Correns, the propagula are produced only on the stem. His efforts to produce them on the leaves by cultivation met with little success, a few threads being sometimes developed on the sheathing base of young leaves, but none on the lamina. When considering *E. procera* he quotes Berggren to the effect that portions of the leaves of this species "readily develope protonema when brought into favorable positions." He concludes with the statement that "of these two nearly related species the monoecious (*E. procera*) is better provided for asexual multiplication than the dioicous (*E. contorta*.)" This seems to be borne out by the specimens from Lockport as the propagula are not confined to the stem as in *E. contorta*.

The moss on Presque Isle was found in fissures on the north face of a cliff or wall of magnesian rock, a dolomite in composition. Those at Lockport grow on the east face of a low cliff of magnesian limestone, dolomitic in character. A little cushion was found in a shallow depression of the vertical face at the base of a tuft of *Pellaea atropurpurea*, another on the edge of the cliff where some soil had accumulated and was associated with *Reboulia hemispherica*: a considerably larger tuft on a narrow ledge with a thin soil mixed with *Brachythecium acuminatum rupincolum*. These plants like the mosses at Presque Isle show the general character of its associates. The habitat is shady but not moist or only slightly so, more commonly dry. Roth gives the habitat of *E. procera* as "shaded schistose rocks." It is usually mentioned as growing in moist places as well as *Myurella Caryana* and *Distichium capillaceum*. But the latter occurs in dry situations also. The associates at Lockport are in general xerophytes, though both Reboulia and the Brachythecium likewise occur as mesophytes.

The discovery of *E. procera* at the two places, Presque Isle and Lockport, south of their usual range, is at first view a little surprising. On the

Eastern Continent it is found in high latitudes, Brotherus (Die Natürlischen Pflanzenfamilien, 1, 3:438, 1902) gives its range as Norway, North Finland and Lapland, Beeven Island and Spitzbergen, also Siberia, near the mouth of the Yenisei River. Several of these stations are north of the Arctic Circle. It appears on the Western Hemisphere within the Arctic Circle at Clavering Island off the coast of Greenland and nearest the two above mentioned island stations north of Europe. On the mainland it has been mostly found in the western part of British America. Macoun gives as its range "Rocks and banks amongst the Rocky Mountains, *Drummond.* Crevices of rocks, Ont.; on wet rocks near Hector and at the 'Gap,' Rocky Mountains; also on the bank of the Columbia River at Revelstoke, B. C., 1890, *Macoun.*" (Catalogue of Canadian Plants, Musci, p. 96, 1892.) It was the moss from Hector, B. C., that I had for comparison (Can. Mus. No. 365). If we follow Brotherus in identifying *E. Selwini* Aust. as *E. procera* Bruch, then Vancouver Island (Victoria at the south end) and some more northerly stations will be included. The station in the above range nearest to Presque Isle is Lake Nepigon, north of Lake Superior, about two hundred and fifty miles away. Its presence on the south shore of Lake Superior would be natural enough in a climate more severe than parts of British Columbia and especially Vancouver Island. It is harder to account for it in northern Illinois, but it is only one of a number of northern plants that find a congenial home near the south end of Lake Michigan. It is the southern limit of range for *Pinus divaricata,* the Jack Pine. Other plants of the vicinity are *Linnaea borealis, Betula papyrifera, Salix adenophylla, Equisetum scirpoides, Aster ptarmicoides lutescens.*

A factor of great weight in accounting for the southern distribution of boreal plants is their relation to former lines of drainage. The locality at Lockport is on the outlet of Lake Michigan where its water, together with those of the upper Lakes, went southward to the Mississippi. It is a glacial made valley through which now flows a small river, the Desplaines. The cliff is vertical, the base covered by thallus, the top forty or more feet higher than the glaciated rock-bed of the valley. Except by the slow process of weathering to which the plants readily conform, the upper part of such a mass would be unchanged since the ice-sheet left it. The lower part would be subject to the wear of the water of the outlet. That plants migrate from the north during the cold of the ice-age is now a common concession. That they withdraw as the ice receded is equally granted, such being left as could adjust themselves to changed climatic conditions. There is no objection, based on climate, to the presence of plants in northern Illinois that are found on the south shore of Lake Superior or in the parts of British Columbia and Vancouver Island near the Pacific Coast, and there is little probability that a moss sheltered as on the rocks at Lockport once established, would be subject to extinction by violence. There is no evidence to show that they were liable to overflow by the former outlet of the lakes. As now constituted there is a valley cut in the limestone at this point forty feet deep and a mile and a quarter wide. Only a very narrow part of it has been deepened by the river that now occupies it, whose shallow bed is but a few feet below the bed of the ancient river, a flat rock bottom covered by a few inches of soil.

Chicago, Ill.

LICHEN NOTES—No. 1.

G. K. MERRILL.

Since the publication of Prof. Fink's paper on *Cladonia verticillata* Hoffm. (BRYOLOGIST. 7:6, 1904) several correspondents have enquired concerning the validity of *C. gracilis* (L.) Nyl. var. *verticillata* Fr. as the name given to plants of similar appearance in their herbaria. To these we will say: the nominations are equivalent and stand for identical forms.

The synonymy of the plant is a varied one. Called by Hoffman (Deutsch. Fl. (1796) p. 122) *C. pyxidata, *C. verticillata*, that name is retained in the works of Floerke, Schaerer, Babington, Coemans, Wainio and others, all of whom concede to it specific rank. Elias Fries (Lich. Eur. Ref. (1831) p. 219) was first to correlate the plant with *C. gracilis*, and it is only natural that our own Tuckerman, pupil of the great master, and throughout life dominated by his teachings, should take the same ground. Such a view is not difficult, as all field workers will agree. We often find undoubted *C. gracilis* proliferous from the center of scyphi, and again verifiable *C. verticillata* with elongated podetia and narrowed cups or rarely ascyphiferous forms with subulate terminations. In our copy of Macoun's Canadian Lichens, No. 295, this condition of *verticillata* is beautifully exemplified and furnishes a remarkable exhibit of transitional tendencies. Nevertheless *C. verticillata* is sufficiently differentiated from *C. gracilis* when examined in typical specimens to be considered distinct. The connecting forms are of no more importance than those serving to link other species in the polymorphous potpourri of the Cladoniaea.

The chemical test with KHO seems to be without particular value in this group. Wainio states that there is no reaction in *C. verticillata*, and in *C. gracilis* there may or may not be. Leighton declares that no reaction is noted with either, while Parrique finds none with *C. gracillis chordalis*, doubtful results with *C. gracilis elongata* and none with *C. verticillata* with the exception of varieties *subcervicornis* Wainio and *Krempelhuberi* Wainio, both of which owe their separation to this feature.

Wainio subdivides the verticillata group into three varieties with several forms and modifications. Comprehending within the meaning of the term variety those forms of closest adherence to the type (varietas constantior=v.), he uses the word form to define phases of perverse development, not self determinative (forma autogenetica inconstans=f.) and modification is applied to anamorphic conditions produced through peculiarities of environment (modificatio inconstans statione producta=m.)

Our continental North American representatives so far known are:

C. verticillata Hoffm. v. evoluta Th. Fr.

v. evoluta m. phyllocephala Flot.

v. evoluta f. apoticta (Ach.) Wainio.
m. cervicornis (Ach.) Flk.
m. abbreviata Wainio.

C. verticillata evoluta Th. Fr. is well described in Prof. Fink's article l. c. under the specific definition of *C. verticillata*, and our notes are intended to be but supplementary. Quoting from Th. M. Fries, Lich. Scan. Pt. I- p. 83, in diagnosis of the species " podetia breviscula " is given as a characteristic. Our American plants are often of very robust habit and considerable height. Podetia has been noted 70 mm. in length. While simple forms are frequently found, this condition if abundant may be taken for abortive. The plants are normally three to five ranked, and eleven were counted in a specimen from Prof. Macoun. The cups are variable in diameter, very narrow in the terminating scyphus and at times reaching 20 mm. with the first rank. Podetia are often found proliferating from the sides as well as from the cups. These are observed to take an initial direction at right angles to that of the podetia, at length if the plant is erect bending to conform. If the parent podetia is deflected the proliferations no longer conform but assume a perpendicular. Instances have been noted where branches originating on the under side of a bent podetia recurved to an upright position. Krempelhuber nominated this phase of development as f. *lateralis* (Lich. Bay. p. 107.). In this connection it may be of interest to record that Schaerer (Enu. p. 195) called those plants proliferous from the center of the cups, f. *centralis*, from the margins, f. *marginalis* and F. *aggregata* (Del.) Malbr. (Supp. Lich. Norm. p. 11) was applied to such as were numerously proliferate from within the scyphus. The variety *evoluta* is well represented by Plate XI, fig. 2, accompanying Prof. Fink's article l. c., and it is difficult to conceive of any other place for Fig. 1 from the point of view developed by Wainio. It may be said for the benefit of the students that (sensu Wainio) the nomination *C. verticillata* per se stands for the entire group. No type form is recognized and the varying phases of the plant are referred to some one of the varieties, forms or modifications or to transitional states between. Prof. Fink's article gives a wide distribution for the variety, we can add nothing to it.

C. verticillata evoluta m. *phyllocephala* Flot. is differentiated principally from v. *evoluta* by the podetia being more or less squamulose, often densely so within the cups. Corticated the same and proliferating similarly its divergence from *evoluta* is controlled by external influences. Symphycarpous states (evidently abortive) are often met with and suggest the idea that similar conditions of other species are likewise abortive. The phenomena of foliolose development in the Cladoniaea is little understood. Two forms are noted, the first (pseudo-foliola) formed by breaking up the cortical tissue into irregular leaf-like expansions. The second (true-squamae) either originating with the podetia, or, initially extrinsic, and communicated through agencies yet undefined from plants normally squamulose to those usually free. Instances have been noted of Cladonia colonies comprising several species in which all were beset with squamae, when only one member of the group was recognized to be typically folioliferous. Specimens have been examined from Massachusetts collected by Miss Carr and Mr. Walter Gerritson. It has also been collected on Mt. Washington.

C. verticillata evoluta f. *apoticta* (Ach.) Wainio. A very curious form collected by Miss C. M. Carr, on sterile soil, Sudbury, Mass., is referred here.

C. verticillata m. *cervicornis* (Ach.) Flk. Examination of numerous European examples emphasize the fact that the caespitose macrophylline primary thallus is the only really distinctive character. Typically inconstant in all other of its exprsssions the Europeans have applied a host of definative phrases to the varying conditions. Such American material as we have seen is referable to Floerk's f. *phyllophora* (De Clad. p. 28) and Nylander's f. *polycarpoides* (Li. Par p. 30.) the former received from Prof. Macoun collected at Mt. Murray, Quebec, and the latter from Miss Carr, collected in Sudbury. Of Miss Carr's plant we can only provisionally place it. European specimens of f. *polycarpoides* are provided with a conspicuous primary thallus, in the Sudbury plant the thallus is deficient. Nylander unites the form with his v. *cervicornis*, but our specimen seems more an expression of v. *evoluta*. It is true that in the obliteration of the scyphi by dissection into fastigiate branchlets there is a strong point of resemblance to cervicornis but ours occurred unmixed with V. evoluta, and may be but a modification.

C. verticillata m. *abbreviata* Wainio, constituted on material collected by Henry Willey in New Bedford, Mass., seems on the whole to be but an expression of m. *cervicornis*. Reduced in both thallus and podetia the state is probably the result of aridity of habitat. Examination of a French specimen detetmined by Wainio discloses many points of similarity with m. *cervicornis*, the most conspicuous of which was the congested horizontal thallus. We have recently found the plant in our own region.

Rockland, Maine.

A NOTE ON LOCAL MOSS DISTRIBUTION.

John M. Holzinger.

The mode of occurrence of several mosses characteristic of the vicinity of Winona, Minn., on the bluffs facing the upper Mississippi valley, appears to follow a very definite law of distribution on a small scale. This fact did not become clear to the writer till this summer, when an opportunity was afforded of a somewhat close exploration for mosses of the bluffs around Dakota, a village on the banks of the "Great Father of Waters," some twenty miles below Winona. Briefly stated, the characteristic mosses near river level (820 feet above the Gulf of Mexico) are *Barbula obtusifolius, Bryum pendulum* and *Leptobryum pryiforme*, all occurring in great abundance on perpendicular sand ledges kept moist more or less throughout the year, and lying in the shade, The Chicago, Milwaukee and St. Paul Railroad here skirts the west bank of the river from La Crosse to St. Paul. Its bed runs just above high water mark. In many places it is hewn out of the solid sand rock. The perpendicular ledges thus exposed are much of the day in the shade, and wherever water trickles from above they become covered with an interesting and uniform moss vegetation, the species mentioned being the most characteristic. Another companion forming extensive dense mats and fruiting abundantly, is *Gymnostomum curvirostre*, but this and the *Bryum* seem able to stand a somewhat drier situation, for depauperate forms of both occur also higher up on the bluffs.

In striking contrast with this river level flora is that at the top of the bluffs, only four hundred feet higher. The most abundant and most characteristic moss here is *Grimmia teretinervis*, covering the calcareous sunburnt sand rock, exposed to an all day sun, in black flat cushions, which are hardly recognized as plants by young students. This species is always sterile. On the same level occur also, with the same regularity, only less abundantly, *Coscinodon Kaui* and *C. Wrightii*, both fruiting. Their principal companions are *Tortula ruralis, Tortella fragilis* and *Ditrichum flexicaule brevifolium*, all sterile.

We have thus on a small scale a real arid region moss flora at the tops of our Mississippi bluffs enjoying essentially the conditions prevailing in dry countries. While at the base, near the river level, there thrives another uniform flora characteristic of better watered regions. And the interesting part is the fact that this narrow strip in altitude extends for fully a hundred miles or more along our Great River.

This peculiar parallelism is made possible by the geological formation, the several alternating strata of Lower Silurian lime and sand rock lie in the Upper Mississippi Basin nearly as horizontal as when they were laid down millions of years ago, so that the Great River with its tributaries large and small, has made its valley by erosion through these strata establishing identical conditions at the two levels referred to.   Winona, Minn.

DIE EUROPAEISCHEN LAUBMOOSE.
By G. Roth, Leipzig, Verlag von Wilhelm Engelmann.
JOHN M. HOLZINGER.

This bryological work comprises two 8vo. volumes of considerable size and appeared in parts in rather rapid succession since early in 1904. The first volume comprises the Cleistocarpi and Acrocarpi to Bryaceae, and covers five hundred and ninety-eight pages of printed matter, exclusive of thirteen pages of Table of Contents, and fifty-two plates closely crowded with drawings of microscopic details of the species described. The second volume, completed in 1905, concludes the Acrocarpi and completes the Pleurocarpi, covering seven hundred and thirty-three pages, accompanied by sixty-two plates similarly replete with figures. The descriptions are accurately drawn up and go into details after the fashion of Limpricht, supplementing in a very helpful way the short and often too incomplete descriptions of many older species. Great care is taken to record the substratum, mode of occurrence and the distribution of the species treated. The legend for the figures illustrating the species follows each description and always includes the record of the exact specimen used in making the drawings, a practice deserving to be universally adopted.

From the systematic view point this effort of Mr. Roth rounds out and brings up to date in accessible form for use, the achievements on European mosses since Schimper's monumental task on his Bryologia Europaea. The plates are necessarily crowded and are largely restricted to drawings of microscopic characters, so as to bring the work within the compass of reasonable accessibibility.

The noteworthy feature of Mr. Roth's work is not so much in his contribution to systematic Bryology, which is considerable, as it is in his primary motive, steadfastly followed throughout, of tracing the intimate relation of the moss flora of the earth to the general economy of nature, their influence on and relation to the soil, to moisture and especially to forests. This fact will stand out more strikingly after a careful perusal of the general discussion introducing the subject, covering pages one to ninety-two of the first volume.

In this introductory treatment the author disposes first of the anatomical structure of the mosses, presenting in well sustained discussions the facts of protonema, moss stem, moss leaf, inflorescence, sporogone, propagation and distribution. And so far he is in full accord with the treatment of the subject in other comprehensive works on mosses. It is, however, in the closing essay on the general subject, covering pages sixty-two to seventy-eight, that Mr. Roth departs conspicuously from the single purpose of the systematist, in that he shows convincingly the importance and the significance of mosses in the economy of nature in agriculture, and especially in forestry. Something of a fair conception of his claims for mosses may be gained from the twelve theses he expounds in this striking essay; 1. Mosses diminish the danger of inundation. 2. They hinder gullying of the soil and aid in the steady flow of springs by increasing the quantity of water derived from condensation and seepage. 3. They preserve the porosity of the soil. 4. They maintain and increase the humidity of the soil. 5. They aid in the formation of humus and so increase the depth of good soil. 6. They introduce the decay of rocks. 7. They effect an equalization of the temperature of the soil. 8. They may be used as bedding material by the agriculturist. 9. They constitute a useful index of conditions in the formation and improvement of meadows. 10. They likewise furnish to the forester a helpful index of climatic conditions favorable to different forest trees. The treatment of this point closes with these words, " If the forests officials who in recent years traveled all over North America for the purpose of studying the then existing forests had only brought back with them the principal representatives of the moss vegetation found in the forest regions traveled by them, we would be much more easily able to form a judgment concerning the adaptability of the several forest trees to our (German) conditions " 11. Mosses protect forest trees against the effects of too great cold. 12. On the part of man some mosses are used in the industries and in the domestic life.

While the author thus fairly exhausts the various relations of mosses to man and nature, it is worthy of note, as he confesses in the first lines of his Vorwort, that in his own case this study was primarily taken up, and reached its full proportions principally because of his conviction that a knowledge of mosses must be of great value to both the agriculturist and especially the forester. And it was his desire to facilitate their study by the practical foresters of Europe that led him in the first instance to essay the publication of his work, a motive that has given so distinctive a coloring to these two volumes.

Worthy of mention here is also the Index of the Literature of European Mosses, covering pages ninety-three to one hundred of Volume I, a supplement to which appears on pages fifteen and sixteen of Volume II, bringing all references up to date, 1905.

The writer of this notice has only one regret to record; it is due to the absense of dichotomous keys, now found in all such treatises, from this otherwise superior work. The insertion of such keys with at least the larger genera would make the books much more usable, but even without them, Mr. Roth's descriptions are very valuable to the systematic bryologist because of the microscopic details recorded. Both descriptions and drawings are a lasting monument to the ability and diligence of the author. It is to be hoped that he will also publish his work on exotic mosses, of which he states he has already drawn over two thousand eight hundred and fifty species. Winona, Minn.

SULLIVANT MOSS CHAPTER NOTES.

The following names have been added to the list of Chapter Members since September 1st, making the total number one hundred and fifty-six: Mr. Reginald Heber Howe, Jr., Middlesex School, Concord, Mass.; Prof. H. A. Green, Tryon, North Carolina; Dr. H. E. Hasse, Soldiers' Home, Los Angeles Co., Calif.

Mr. G. K. Merrill, 564 Main street, Rockland, Me., will be glad to determine any and all species of lichens sent him provided the specimens are ample and are accompanied by full data.

NOTICE—ELECTION OF OFFICERS FOR 1906.

Please forward your ballots AT ONCE to the Judge of Elections, Miss Cora H. Clarke, 91 Mt. Vernon street, Boston, Mass. Polls close November 30th.

For President — Mr. Edward B. Chamberlain, 1830 Jefferson Place, Washington, D. C.

For Vice President — Mr. G. K. Merrill, 564 Main street, Rockland, Maine.

For Secretary—Dr. John W. Bailey, Walker Building, Seattle, Wash.

For Treasurer—Mrs. Annie Morrill Smith, 78 Orange street, Brooklyn, N. Y.

OFFERINGS.

(To Chapter Members only. For postage.)

Miss Harriet Wheeler, Chatham, Columbia Co. N. Y. *Orthotrichum anomalum* Hedw. c.fr. Collected Queechy Lake, N. Y. *Hylocomium umbratum* B. & S., st. Collected in White Mts., N. H.

Mr. John A. Anderson, High School, Dubuque, Iowa. *Climacium Americanum* Brid. c.fr. Collected in Iowa.

Mr. B. D. Gilbert, Clayville, Oneida Co., N. Y. *Brachythecium plumosum* (Sw.) B. & S. c.fr. Collected in Clayville.

Mr. A. S. Foster, 282½ Second street, Portland, Oregon. *Scouleria marginata* E. G. Britton, c.fr.: *Pogonatum urnigerum* (L.) Beauv., c.fr,; *Alsia abietina* Sulliv. Collected in Oregon.

Miss Cora H. Clarke, 91 Mt. Vernon street, Boston, Mass. *Hypnum cordifolium* Hedw. c.fr. Collected in Mass.

Mr. C. C. Plitt, 1706 Hanover strreet, Baltimore, Md. *Hypnum cordifolium* Hedw. c.fr. Collected in Maryland.

Miss Mary F. Miller, 1109 M. street, N. W., Washington. D. C. *Pogonatum contortum* Lesq. c.fr.; *Dicranoweisia cirrhata* Lindb. c.fr. Collected in British Columbia by Mr. A. J. Hill.

Miss Annie Lorenz, 96 Garden street, Hartford. Conn. *Sphagnum sedoides* Brid. st.; *Mylia Taylori* (Hook.) S. P. Gray. Collected on Mt. Marcy, N. Y.

Miss Caroline C. Haynes, 16 East 36th street, N. Y. City. *Nardia crenulata* (Smith) Lindb.; *Telaranea nematodes longifolia* M. A. Howe, Collected in New Jersey.

Mrs. Sarah B. Hadley, R. F. D. No. 1, South Canterbury, Conn. *Peltigera canina* (L.) Hoffm.; *Umbilicaria Muhlenbergii* (Ach.) Tuck.: *U. pustulata* (L.) Hoffm. var. *papulosa* Tuck. Collected in Conn.